Conservation of Plastics

This book is dedicated to the memory of my much-loved,
Maureen Shashoua, who was fascinated by plastic jewellery and
designs and inspired me greatly.

Conservation of Plastics

Materials science, degradation and preservation

Yvonne Shashoua

Routledge
Taylor & Francis Group

LONDON AND NEW YORK

First published 2008 by Butterworth-Heinemann

2 Park Square, Milton Park, Abingdon, Oxfordshire OX14 4RN
52 Vanderbilt Avenue, New York, NY 10017

Routledge is an imprint of the Taylor & Francis Group, an informa business

First issued in paperback 2020

British Library Cataloguing in Publication Data
A catalogue record for this book is available from the British Library

Library of Congress Cataloging in Publication Data
A catalog record for this book is available from the Library of Congress

ISBN: 978-0-7506-6495-0 (hbk)
ISBN: 978-0-367-60630-5 (pbk)

Typeset by Charon Tec Ltd., A Macmillan Company. (www.macmillansolutions.com)

CONTENTS

FOREWORD

Plastics are the materials of today. It is hard to imagine a modern-day product that does not use plastics. In countless cases, products which we take for granted could not have been developed without plastics. From toothbrushes to telephones, from computers to cars, from electrical appliances to aeroplanes, plastics are the wonder materials of this modern age. Increasing amounts of plastics are also being produced and used by artists. Plastics are thus making our lives comfortable, enjoyable and safe.

The need for knowledge about the conservation of plastics in cultural heritage is thus becoming increasingly important. Although a very substantial body of literature exists on the chemical, technological and manufacturing aspects of plastics, the scope of such books is too broad for conservation needs. Moreover, they often go into far too much depth to be directly useful for conservation purposes. On the other hand, the books which have been published until now about plastics conservation deal only with one or two plastic materials, and thus do not provide a total overview.

The first time I wished for a book like *Conservation of Plastics* was in 1990 when I started doing research on the conservation of cultural heritage plastics. At that time conservators could only use the introduction sections on modern materials conservation from conference proceedings published in the late 1980s and early 1990s as references. In 1993 *Saving the Twentieth Century* was published. It comprised the postprints of the conference of the same title held in 1991. *Saving the Twentieth Century* described exactly the problems which I encountered during my first years in plastics research. In 1997 the conference postprints *Modern art, who cares?* was published, providing good insight into the problems of modern art conservation. However, more information about plastics materials was needed.

There has long been a strong need felt by the conservation community for a book that provides a complete understanding of plastics degradation and conservation and that covers the latest advances in technology. This publication, *Conservation of Plastics*, is exactly what the field needs. It fills the gap between books describing the chemistry and technology of plastics on the one hand, and the traditional conservation literature on the other. Moreover, it facilitates the understanding of the relationships between the chemical and physical background of plastics, degradation processes, conservation treatments, and the behaviour of objects during those treatments. With its broad scope, it is likely to benefit collectors, conservators, curators, conservation scientists, and anyone with interest or responsibility for the care of plastics.

The Conservation of Plastics provides important information on the definition of plastics, their history, technology, properties, identification, degradation, conservation and future preservation. Most importantly, it describes the state of the art of one of the newest fields in cultural heritage – plastics conservation. The information in this book is placed in context, and is illustrated with many examples that those involved in decision-making processes for conserving plastics objects will find helpful. This book will be present on every conservation table, in all museums, training schools and universities, where it will be used on a daily basis. The beneficiaries will be the objects themselves: plastics for everyday use, designer objects, furniture, toys and, most of all, modern and contemporary art.

The author has made a tremendous effort in preparing this book, focussing attention on both basic and advanced aspects. Yet, she has kept it simple and readable, making it an enjoyable read which will help any beginner and expert alike to understand plastics conservation. This book has been very thoughtfully divided into eight chapters, each of which has a logical sequence of sections. It includes the latest trends, where the emphasis has been laid on the future needs of plastics conservation. This book reflects the author's rich experience in more than 15 years of interdisciplinary and international activities in academia, industry and conservation science.

It is not often that one gets a chance to write a foreword to a book which has immense potential to become a standard handbook for the conservation of plastics in cultural heritage. I congratulate Yvonne Shashoua for this fine contribution, and am happy to recommend this pioneering book to all in the field.

Thea B. van Oosten
Senior conservation scientist
Netherlands Institute for Cultural Heritage (ICN),
The Netherlands, Amsterdam
Former coordinator Modern Materials and
Contemporary Art Working Group ICOM-CC
Member of the board of ICOM-CC

PREFACE

Plastics have had a significant influence on industrial, domestic and cultural aspects of everyday life in the twentieth and twenty-first centuries. They represent advances in technology, illustrated by the dramatic growth in number and type of information storage media available since the 1970s, credit and payment cards, medical applications and food containers which can be taken directly from freezer to microwave oven to dinner table. The development of plastics reflects economic history. Restrictions on imported rubber latex, wool, silk and other natural materials to Europe during World War II stimulated the development of synthetic alternatives. Plastics may also be collected to study a society's attitudes towards new materials and technologies. Plastics are increasingly used to create artworks. In short, plastic's significance is reflected in the prophecy made by artists Marcel Biefer and Beat Zgraggen in 1991 which states that 'plastic artefacts will be the most important witnesses to our time'. While museums continue to acquire objects which reflect everyday life and historical events, the proportion of plastics in collections will increase.

Plastics have short lifetimes compared with those of traditional materials found in museums, exhibiting deterioration within 5 to 35 years of acquisition. Deterioration and conservation of plastics objects in museums has only been recognized formally as an area worthy of research since 1991. However, despite its novelty, the conservation of plastics is a rapidly developing field. A measure of the recognition of plastics conservation as a serious discipline was the creation of the working group 'Modern Materials and Contemporary Art' (MMCa WG) by the International Council of Museums Committee for Conservation in 1996.

Museums and other cultural institutions aim to conserve plastics for future generations and to preserve information about outdated materials and technologies. In addition to conserving the plastic material itself, preserving the design, concept or intention reflected by the object is often important. Private collectors wish to preserve the artistic, historical and financial values of their objects and artworks. With the dramatic increase in monetary value of modern art since 2000, art insurance companies are also concerned with the short useful lifetime of plastics.

The purpose of *Conservation of Plastics* is to distil the extensive knowledge produced by the polymer and plastics industries, designers, environmental and conservation professionals into a single publication focussing on the preservation and conservation of plastics. The book focusses on three-dimensional objects constructed from semi-synthetic and synthetic plastics. The book progresses from the historical development of plastics, through production and the chemical and physical properties of the materials formed to identification and the factors causing physical, chemical and biological degradation. Tables of optical, physical and chemical properties of the most common plastics in collections, one for each material, are presented in Appendix 1 to assist both selection and identification of plastics. Photographs of flame tests used to identify plastics augment written descriptions. Terms used to define degradation are illustrated with photographs in Appendix 3. *Conservation of Plastics* describes and discusses critically the developments in techniques and materials used to conserve semi-synthetic and synthetic plastics and proposes tools to further develop this new discipline. Future directions for conservation of plastics are also proposed.

ACKNOWLEDGEMENTS

It would not have been possible for me to write this book without the support of my employer, the National Museum of Denmark. I was given a paid research year to review the literature and prepare the text, which expanded into 20 months. I am extremely grateful to Jesper Stub Johnsen, Head of Conservation at the National Museum of Denmark and Mads Christian Christensen, Head of the Research, Analysis and Consultancy section, who took an active interest in the book and were always positive about its progress. My colleagues in the Conservation Department, particularly Poul Jensen, Jens Glastrup and Martin Mortensen, have been generous with their time to discuss aspects of the text and have inspired and supported my work from start to finish. I am also grateful to Roberto Fortuna and Kira Kroeis Ursem, photographers in the Conservation Department, for their creativity, particularly in capturing images of burning plastics and librarian Pla Olsen for her effectiveness in obtaining references at short notice.

I received invaluable comments on all chapters from Brenda Keneghan, polymer scientist at the Victoria & Albert Museum and Louise Cone, conservator at the Danish National Gallery. Both gave up their free time, sometimes at short notice and always without complaint. The quality of the information and ideas contained in the book is higher for their experienced input. Anita Quye, conservation scientist at the National Museums of Scotland, and Thea van Oosten, senior researcher at the Netherlands Institute for Cultural Heritage, contributed greatly to the book at the planning stage. I have also learned much from the comments on style given by Tim Padfield, consultant in museum climate.

On a personal note, I wish to thank Jan K. Madsen, Ron, Neil and Michelle Shashoua for their constant patience, support and belief in my ability to complete this book.

PICTURE CREDITS

Chapter 1

Figure 1.2: Reproduced by kind permission of Fiona Hall; © National Gallery of Victoria

Figure 1.3: © Pamela Wells

Figure 1.4: Reproduced by kind permission of Miwa Koizumi, PET project, 2005; © Miwa Koizumi

Chapter 2

Figure 2.1: Photograph by John Lee; © National Museum of Denmark

Figures 2.3, 2.8, 2.11 and 2.12: Photographs by Roberto Fortuna; © National Museum of Denmark

Figures 2.1, 2.4, 2.6 and 2.7: Photographs by John Lee; © National Museum of Denmark

Chapter 3

Figure 3.13: © National Museum of Denmark

Figure 3.16: Private collection. Photograph by Ronald Shashoua

Figures 3.17, 3.18, 3.25, 3.30, 3.34, 3.36, 3.39, 3.40: Photographs by Roberto Fortuna; © National Museum of Denmark

Figures 3.20, 3.21, 3.23, 3.28: Photographs by John Lee; © National Museum of Denmark

Figure 3.32: Reproduced by kind permission of Tom Dixon; © Tom Dixon

Figure 3.41: Photograph by Mika Friman

Figure 3.42: Photograph by Paivi Kyllonen

Figure 3.43: Reproduced with kind permission of Peter Chang; © Peter Chang

Chapter 4

Figures 4.4, 4.5, 4.9: Photographs by Roberto Fortuna; © National Museum of Denmark

Chapter 5

Figures 5.1, 5.20, 5.22: Photographs by John Lee; © National Museum of Denmark

Figures 5.2, 5.3, 5.4, 5.5, 5.6., 5.7, 5.8, 5.9, 5.12: Photographs by Roberto Fortuna; © National Museum of Denmark

Chapter 6

Figures 6.1, 6.4, 6.7, 6.11, 6.13, 6.17, 6.19, 6.22, 6.28: Photographs by Roberto Fortuna; © National Museum of Denmark

Figures 6.2, 6.3, 6.9, 6.26: Photographs by John Lee; © National Museum of Denmark

Chapter 7

Figures 7.3, 7.4, 7.5, 7.9: Photographs by Roberto Fortuna; © National Museum of Denmark

Figure 7.8: Reproduced by kind permission of Antonio Rava restoration workshop; © National Museum of Cinema, Torino

Figure 7.12: Photograph by Ronald Shashoua

Appendix 3

All photographs by Roberto Fortuna; © National Museum of Denmark

Plastics in collections

Summary

Chapter 1 defines the term 'plastics' as used in the book (only 3D materials will be addressed, not paintings or synthetic lacquers or photographic materials). The acquisition, significance, condition and current status of plastics conservation are discussed.

Plastics are semi-synthetic or synthetic materials which can be manipulated to form films, fibres, foams or three-dimensional objects. Natural polymers from plants, insects and animals are not discussed in this book. Synthetic paints, synthetic textiles and photographic film contain plastics but will not be discussed in this book because their conservation is the focus of specialist publications (Chiantore and Rava, 2005; Learner, 2005; Tímár-Balázsy and Eastop, 1998; Lavedrine et al., 2003). Plastics are based on polymers, also known as macromolecules, which are large molecules made by joining together many smaller ones. The chemical and physical properties of liquid polymers are modified with additives and shaped to convert them into solids with dimensionally stable forms.

Increased numbers of processing and fabrication techniques have allowed modern plastics to be manipulated in thin film, bulk and foam forms, and to be combined and reinforced with fibres, metals, wood and other materials. Different plastics types can also be laminated together using heat. Today there are approximately 50 different basic types of polymer included in 60 000 plastics formulations. The world's annual consumption of plastic materials has increased from around 5 million tonnes in the 1950s to nearly 100 million tonnes in the year 2000. In 1982, plastics production surpassed that of steel worldwide and that year has been heralded as the start of the Plastic Age (Ward, 1997). Packaging is the biggest market sector for plastics worldwide (Figure 1.1). Polyethylene and poly (vinyl chloride) are the most highly consumed of all plastics worldwide (Brydson, 1999).

Figure 1.1

Applications of plastics worldwide (Morgan, 1994).

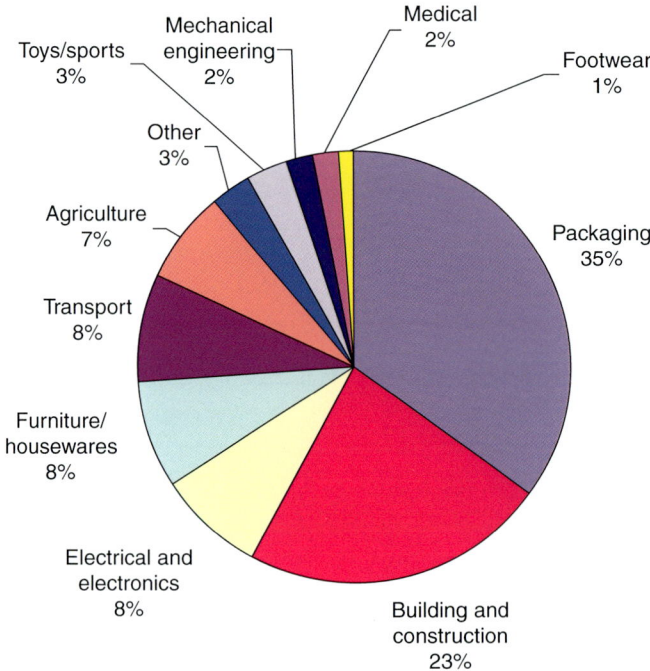

The rapid growth in plastic applications during the twentieth century may be attributed to the properties of plastics, which include:

- high versatility and ability to be tailored to specific technical needs
- little or no finishing, painting or polishing required
- lower density than competing materials, reducing fuel consumption during transportation
- excellent thermal and electrical insulation properties
- good safety and hygiene properties for food packaging.

Because the raw materials for plastics are based on crude oil, there is close correlation between their prices. Although plastics are generally considered to be low cost raw materials, at around £1500 per tonne, they compare poorly with their main competitors including steel at £250 per tonne, and aluminium and zinc at £1000 per tonne.

1.1   Collecting plastics

Museum objects are rarely collected for their material type alone but because of their origin, function, design, rarity, cultural, historical or artistic significance. Plastics are collected for the same reasons. Synthetic plastics have had a significant influence on industrial, domestic and cultural aspects of everyday life in the twentieth and twenty-first centuries. Their significance is reflected in the prophecy by artists Marcel Biefer and Beat Zgraggen in 1991 which states that 'plastic artefacts will be the most important witnesses to our time' (Biefer and Zgraggen, 1991).

Plastics represent advances in technology, illustrated by the dramatic growth in the number and type of information storage media available since the 1970s, credit and payment cards and food containers which can be taken directly from freezer to microwave oven to dinner table without breaking. Before the 1940s, it was not possible to drink hot coffee from a plastic cup without it softening and becoming too hot to hold – an activity which is commonplace today. Plastics are also collected to show the social effects of technological development.

Industrial archaeology is a rapidly expanding, interdisciplinary method of studying material and immaterial evidence created for, or by, industrial processes from the beginning of the Industrial Revolution in the second half of the eighteenth century up to and including the present day. Traditionally, industrial heritage has concerned mines, steelworks and other heavy industries, but recent projects include the study of plastics. In addition to allowing the study of technical processes, collecting plastics provides associations with living conditions of the plastic workers, their health and level of education. In 1995, the Norwegian Museum of Science and Technology initiated a collaborative project known as 'Plastics in modern Norway'. Two of the project's goals are to show how the use of plastics affects artefacts and consumer culture in Norway today and to stimulate collaborative research and collection of contemporary history (Rossnes et al., 2002).

Plastics may also be collected to study a society's attitudes towards new materials and technologies. When the first man-made plastic, cellulose nitrate, was exhibited at the Great International Exhibition in England in 1862 by Alexander Parkes, it was designed to imitate luxury materials, such as tortoiseshell and ivory, which were in increasing demand and diminishing supply. However, the image of plastics as highly valued luxury goods faded when colourful, post-World War II designs were marketed in large numbers and at low cost. Plastics developed a long-lasting image as low value, poor quality and ephemeral pieces. It has even been said that dreaming about plastics suggests that one is fake and artificial! The 1980s saw a change in perception of plastics from disposable materials to fashionable, highly collectable pieces with historical and technological significance. In the twenty-first century, industrial plastics have gained a reputation as pollutants, both during use and on disposal, and as health hazards, particularly to young children. By contrast, art made from recycled plastics is highly collectable today.

The development of plastics reflects economic history. Restrictions on imported rubber latex, wool, silk and other natural materials to Europe during World War II stimulated the development of synthetic alternatives. Between 1935 and 1945, many new polymers were introduced including polyethylene, polyamides, poly (methyl methacrylate), polyurethanes, poly (vinyl chloride), silicones, epoxies, polytetrafluoroethylene and polystyrene. Polyethylene was incorporated into radar systems while poly (vinyl chloride) replaced the limited stocks of natural rubber as cable insulation.

Plastics are increasingly used to create artworks. Such artworks are collected for the concept or meaning they illustrate as well as for the materials used. The first artists to express their artistic ideas using plastics were the Russian-born brothers Antoine Pevsner and Naum Gabo. They regarded plastics as modern technological materials and in 1910 crafted faces and other three-dimensional forms in a Cubist-inspired style using cellulose nitrate sheets. Artists experienced the instability of cellulose nitrate and exchanged it for the more stable Plexiglas or Perspex (poly [methyl methacrylate]) in the mid 1930s. The new plastic had visual properties similar to cellulose nitrate but retained them for longer, and could be worked with woodworking tools and shaped with heat.

The number of plastics used by artists had increased dramatically by the end of the 1960s, despite the growing evidence that plastics were unstable and required either conservation or replacement. The future of the Guggenheim Museum's large sculpture 'Expanded Expansion' by Eva Hesse (1936–1970) was the subject of a one-day meeting in 2000 (Hochfield, 2002). The large sculpture was constructed from latex-soaked cheesecloth panels attached to fibreglass poles. The piece had not been exhibited upright, as the artist intended, since 1988 because of the risk of the brittle panels detaching from the poles *Expanded Expansion* had barely outlasted the life of its artist.

Eva Hesse later employed the translucent, flowing properties of glass-reinforced polyester to represent alienation and processes in subsequent artworks. The American sculptor Duane Hanson (1925–1996) worked with glass-reinforced polyester from 1967 onwards to make very lifelike figures, shown frozen in action, cast from living models and accessorized with real wigs and clothing. The Australian artist, Ron Mueck, also creates lifelike figures but in larger than life format. Like Hanson, Mueck uses glass-reinforced polyester to form the body, but substitutes it with silicon to create the softer facial skin (Albus et al., 2007). Another Australian artist, Fiona Hall, manipulates rigid PVC water pipes to induce a soft, living quality in her sculptures (Figure 1.2a and b).

Although artists throughout history have reused materials and applied discarded objects to create artworks, a form of creativity known as environmental or recycled art was applied extensively to plastics in the late 1990s. Plastic carrier bags, packaging film and plastic water bottles are collected from rubbish bins, dumps, beaches and streets and used as raw materials for collages and sculptures (Figure 1.3). Environmental artists either express a message concerning the Earth's dwindling natural resources and our misuse of the raw materials we have, or they explore the changing role of reused materials. John Dahlsen, an Australian artist, uses recycled carrier bags as his primary medium and imagines them to be both on the verge of extinction due to governmental

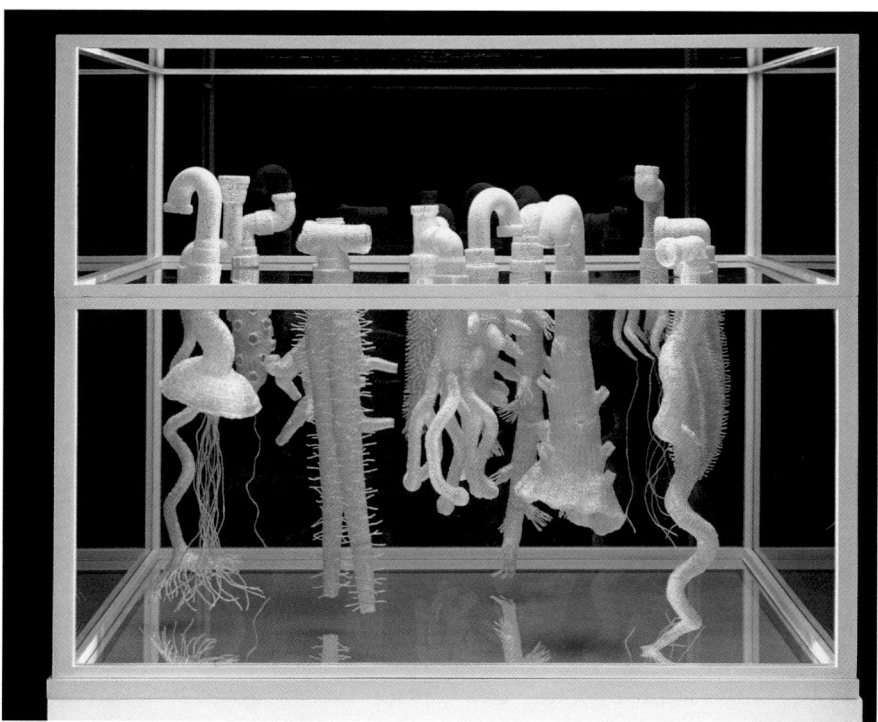

Figure 1.2a
Dead in the water
by Fiona Hall is
constructed from rigid
PVC water pipes treated
to impart a soft, living
quality.

taxes and a scourge on our environment on a worldwide scale (Dahlsen, 2007). Miwa Koizumi, an artist living in New York, cuts and thermoforms poly (ethylene terephthalate) (PET) water bottles to form jellyfish and anemone as part of her PET Project works (Figure 1.4). Koizumi explains that the water bottle changes status from trash to non-trash (Koizumi, 2005).

Private collectors and dealers may also acquire plastics for their commercial value. The prices of contemporary art, designs in plastic, jewellery and plastic models associated with films have increased dramatically since the mid 1980s. At auction in February 2007, Naum Gabo's *Linear Construction in Space No. 3* was sold for a record £1 252 000 (The Associated Press, 2007). Previous sales of Gabo's work had realized a maximum £350 500. In September 2006, a Barbie doll manufactured in 1955, 'Barbie in midnight red', was auctioned for £9000 and is the most expensive doll sold at auction. The previous auction record for Barbie dolls was £2800 in 1999. In its original packaging from the 1970s, vinyl models of the character Jawa from the Star Wars films attained up to £2 2129 in 2006. One negative effect of the increase in commercial value has been that jewellery made from copies of Bakelite (phenol formaldehyde), known as Fakelite, appeared for sale in 1988. Fakelite is either Bakelite which has been reworked into another form, such as buttons being adhered to form a brooch, or other plastics which are being falsely described as Bakelite to inflate their value. A website has been set up to warn collectors of the presence of Fakelite (Fakelite.com, 2005).

Figure 1.2b
The apparent fragility is seen on close inspection.

Figure 1.3
Rainbow 13 from Chasing Rainbows – This Beautiful Impossible Persuit by Pamela Wells. This recycled artwork is constructed from stitched polythene collected in Warsaw, Poland.

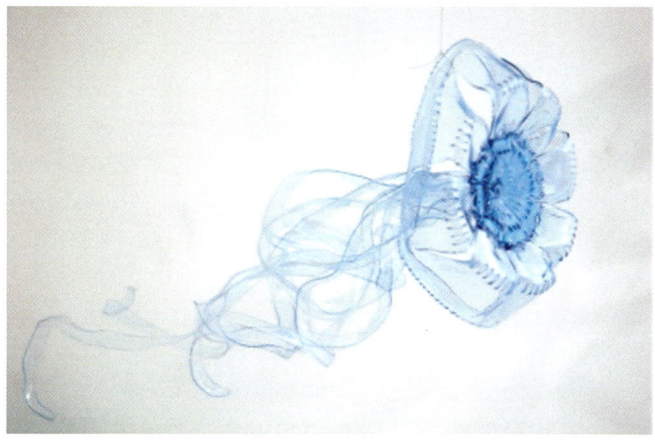

Figure 1.4
Miwa Koizumi's PET
Project jellyfish is made
from a blue Volvic®
water bottle.

Table 1.1 Examples of collections and objects containing plastics	
Type of collection	**Examples of objects**
information carriers	plastic overhead and tracing sheets, maps, tapes, vinyl records, floppy disks, CDs, DVDs, photographic film, plastic newspapers
technology	electronic circuit boards, cable insulation, housings for electrical equipment
transport	upholstery, bumpers and dashboards in cars and lorries, agricultural machinery, bicycles, windows in side cars, fibreglass boats, protective clothing, spacesuits
military	uniforms, protective equipment, rucksacks, tents, parachutes, transport
building	gutters, pipes, flooring, wall coverings, furniture
modern history	clothes, shoes and other fashion accessories such as bags and jewellery, makeup containers, furniture, toys
medical equipment	blood bags, prosthetic limbs and joints, false teeth, spectacles, tubing, syringes
sports equipment	surfboards, racquets, clothing
modern (from 1880s to 1970s) and contemporary art (from 1970s to present)	paintings, collage, sculptures, video film art, photographic art
coins and medals	tokens, badges
packaging	bubble wrap, photograph pockets, laminating sheets, book coverings, plastic bags

Today, almost all international museums and galleries possess collections which contain plastics, either as objects in their own right or as components of composite objects (Table 1.1). The majority of private collections are devoted exclusively to particular plastics, such as Bakelite, or contain a significant proportion of plastics (e.g. button, toy and radio collections), whereas in museum collections plastics are more widely distributed and include components in addition to whole objects of plastic (Morgan, 1994). A study of the types of plastics present in museum collections in the British Museum and the V&A Museum in London in the early 1990s identified phenol- and urea-formaldehydes, cellulose acetate and nitrate, epoxy resins, polyamides, poly (vinyl chloride), polyesters, polyethylene, poly (methyl methacrylate), polystyrene and polyurethanes (Shashoua and Ward, 1995; Then and Oakley, 1993).

The total number of plastics currently in museum and private collections worldwide has not been published. However, a total of 3032 plastic-containing objects was identified in the collections of the British Museum in 1995 from the electronic documentation system in use at that time (Shashoua and Ward, 1995). The V&A Museum in London identified 4500 plastics objects in 1993 (Then and Oakley, 1993). The Science Museum in London had 1500 plastics objects in their collections in 1990 (Mossman, 1993). A survey initiated in 2005 to examine plastics in 51 Swedish museums identified a total of 360 objects in eight museum collections investigated within the first year of the project (Nord et al., 2006). It is difficult to ascertain precisely how many plastics objects are present in museum or art gallery collections, because many objects are neither recognized nor registered as containing plastics components. Brenda Keneghan, polymer scientist at the V&A Museum in London, diagnosed 'plastics denial syndrome' amongst her curatorial colleagues in 1996 (Keneghan, 1996). Plastics were not recognized in museum objects and there was no appreciation of their instability. By 2005, 'plastics denial syndrome' had been largely cured by a dramatic increase in the number of conferences, publications, educational programmes and exhibitions concerning plastics (Keneghan, 2005).

A survey conducted in the early 1990s of nine private collections and 116 museum collections in the United Kingdom conducted in the early 1990s suggested that most objects containing plastics date from 1920 to 1960, but more than 40 per cent of collections contained objects manufactured after 1980 (Figure 1.5). The proportion of post-1980 plastics is likely to increase with time as museums continue to collect materials which reflect modern life and as the older plastics in collections degrade and are deaccessioned.

1.2 Condition of plastics in collections

Investigations to establish the conservation requirements of plastics-containing collections in the V&A Museum (Then and Oakley, 1993) and the British Museum (Shashoua and Ward, 1995) in the early 1990s suggested that around 1 per cent of objects were actively degrading and required stabilizing, while 12 per cent showed degradation which necessitated strengthening or repair (Figure 1.6). Degradation of a plastic is any change which has adverse effects on its

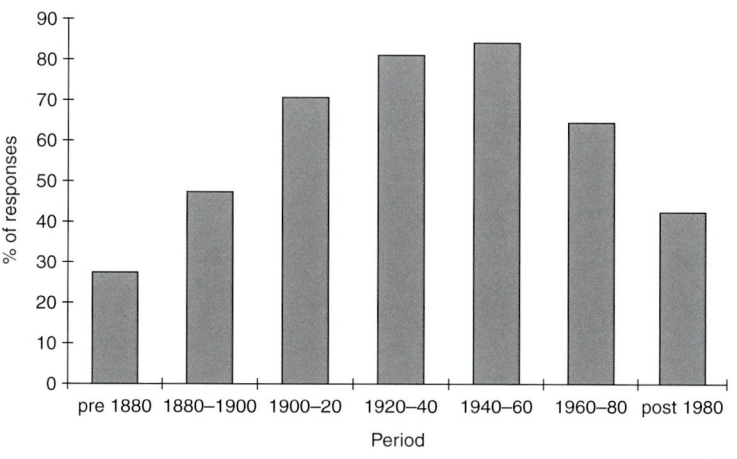

Figure 1.5
Distribution of plastics in museum collections surveyed (percentage of responses by museums) by manufacture period (Morgan, 1994).

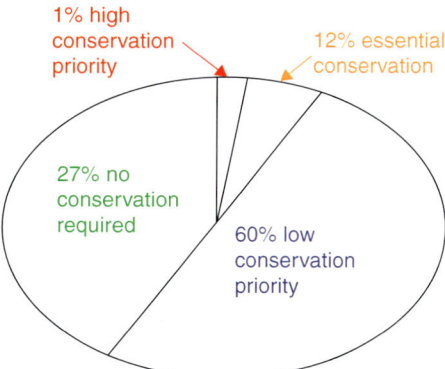

Figure 1.6
Average results of condition surveys of objects containing plastics in UK museums.

properties, function or significance. Most plastics examined (60 per cent) were stable but displayed surface damage such as scratches and dirt and required cleaning. Only around one quarter of the plastics required no conservation.

Because condition surveys demand a high investment in resources, they are not usually conducted at regular intervals and therefore provide only a snapshot of the condition of collections at one particular moment. In general, plastics have an induction period following manufacture, during which no physical or chemical changes can be detected, followed by a period during which irreversible damage can be observed and measured. This period continues until the end of the useful lifetime for the object is reached. The shift from induction to degradation periods is often brief. As a result, plastics which exhibit no signs of degradation and are therefore documented as requiring no conservation can discolour or develop powdery surfaces six months later.

All the unstable objects identified in surveys at the V&A Museum and British Museum contain cellulose nitrate, cellulose acetate, plasticized poly (vinyl chloride) (PVC) or polyurethane foams. The particular instability of these four plastics has also been recognized in surveys of other large collections including ones at the National Museum of Denmark and the National Museum of

Scotland (Quye, 1993). Instability of the earliest plastics, cellulose nitrate and acetate, is expected due to their poorly stabilized and largely experimental formulations and because they are the oldest man-made plastics in museums dating from the late nineteenth and early twentieth centuries. However, PVC and polyurethanes were first commercially available after World War II and are still used, so their short lifetimes are more difficult to understand. Until the late 1970s, plastics were widely believed to last forever and to be almost indestructible, a belief fuelled by the plastics industry's claims for the durability of their products (Morgan, 1994).

Once plastics objects are registered in collections, the owner institution has a responsibility to preserve them for the future, until the end of their useful lifetime, when the object ceases to have either a recognizable physical form or loses its intended meaning (Bradley, 1994). Museums and other cultural institutions aim to preserve objects for the enjoyment and education of the next generation, sometimes defined as a minimum of either 50 or 100 years. The definition of a useful lifetime as applied to cultural materials is different to that traditionally used by the plastics industry, where a predetermined reduction in one or two physical or chemical properties associated with the material's performance is the sole deciding factor. Industrial plastics are designed and formulated to function for a predefined period (Table 1.2).

Table 1.2   Major categories of use and expected lifetimes of PVC products in Europe (European Union Commission, 2000)

Category	Example of application	Average lifetime (years)
building	window frames	10–15
packaging	film and sheet	1
furniture	fake leather upholstery	17
household appliances	tubing	11
electric and electronic	cable insulation	21
automotive	steering wheel cover	12
others	blood bags	2–10

Quackenbos defines the lifetime of a plasticized PVC film as the period required for 10 per cent of its original weight to be lost (Quackenbos, 1954). At that point, the material is considered to have failed its industrial requirements. Extrapolating measurements obtained from accelerated thermal ageing, Quackenbos concludes that lifetimes for plasticized PVC films ranges from 3 months to 1000 years at ambient temperatures, depending on the concentration of plasticizer and thermal stabilizer. Because PVC was first available in the 1920s, materials which have undergone real time ageing for 1000 years are not yet available. However, condition surveys in museums suggest that plasticized PVC exhibits degradation between 5 and 35 years after acquisition (Shashoua, 2001).

Van Oosten argues that before it is possible to accurately estimate the life expectancy of plastics in a museum, it is necessary to establish how much degradation is acceptable before the material shows a loss of quality (van Oosten, 1999). For example, while changes in surface colour, gloss or texture of oil paint, metals and stone are described as patina and valued as a measure of age and authenticity, the same phenomenon on plastics surfaces is regarded negatively as degradation. It should also be considered that the majority of museum objects containing plastics are not new when collected, so may already have degraded prior to collection. Exposure to light, heat, moisture, chemicals and gaseous pollutants during the pre-collection period is likely to reduce longevity.

1.3   Origin and development of conservation of plastics

The reasons for conserving plastics are no different to those for conserving other cultural materials. Museums and other cultural institutions conserve objects for future generations and to preserve information about outdated materials and technologies. In addition to conserving the plastic material itself, preserving the design, concept or intention reflected by the object can be a goal.

Private collectors wish to preserve the artistic, historical and financial values of their objects and artworks. With the dramatic increase in monetary value of modern art since 2000, art insurance companies are also concerned with the short useful lifetime of plastics and the limited options for their conservation today. In January 2003 AXA Art Insurance funded a research laboratory and research into the conservation and restoration of modern plastic furniture at the Vitra Design Museum in Germany (Albus et al., 2007). Artists are also stakeholders in the conservation of their own and their fellow professionals' works. The durability of an artwork, as well as providing a guarantee for the collector, affects the artist's reputation and immortality (Hummelen, 1999). Durability is affected by artists' selection of appropriate materials and techniques. The artist David Hockney is credited with saying 'Love will decide what is kept, and science will decide how it is kept' (Corzo, 1999).

The study of degradation and conservation of plastics in museums has only been recognized formally as an important area worthy of research since the early 1990s (Grattan, 1993), while the conservation of plastics in modern art first became a specialist sub-group of international focus in 1998 (Hummelen and Sillé, 1999). Conservation of plastics is one of the newest disciplines of the conservation profession, and activity and interest is growing. The Plastics Historical Society (PHS) was formed in 1986 and was the first organization in the United Kingdom to draw attention to the heritage of the plastics industry and to celebrate plastics (Plastics Historical Society, 2007). The PHS is an independent society affiliated to the Institute of Materials, Minerals and Mining and is based in London. The Society's aim is to encourage the study of historical plastics and modern polymers in the form of fibres, rubber and elastomers. The Historical Plastics Research Scientists Group (HPRSG) was formed in 1993 with the purpose of establishing the condition of plastics in museum collections and developing collaborative research projects to learn more about their conservation.

A measure of the recognition of plastics conservation as a serious discipline was the creation of the working group 'Modern Materials and Contemporary Art' (MMCa WG) by the International Council of Museums Committee for Conservation in 1996 (ICOM-CC, 2007). HPRSG became a part of the MMCa WG in 2001. The Working Group has had full papers published in the preprints of all ICOM-CC Triennial Meetings since 1996 and organized an interim meeting documented by postprints in 2001 (van Oosten, Shashoua and Waentig, 2002). In 1999 the MMCa WG had approximately 60 members and the number reached 250 in 2006. However, fewer than 20 conservators and conservation scientists are full-time specialists in plastics degradation and conservation in Europe. One consequence of the lack of expertise in relation to the growing number of degrading plastics is that progress in the field of plastics conservation is slow.

Concern for the longevity of plastics and the lack of established conservation techniques have initiated an increasing number of conferences and publications discussing modern materials and modern art. The conference 'Saving the twentieth century: the conservation of modern materials' held at the Canadian Conservation Institute in Canada in 1990 provided the first comprehensive overview of the extent of degradation of plastics in museum collections and the factors causing their breakdown (Grattan, 1993). 'Polymers in Conservation' held in 1991 illustrated the many applications of plastics in museum collections including photographic film, surface coatings and synthetic textiles (Allen et al., 1992). Participants in 'Resins – ancient and modern' in 1995 concluded that some plastics, notably cellulose nitrate, acetates and rubbers, were less stable than others and techniques to slow their rate of degradation were proposed (Wright and Townsend, 1995). Conservators and conservation scientists were the professionals most concerned with the useful lifetime of plastics in museums until the mid 1990s. As a result, pre-1990 publications focussed almost exclusively on the chemical and physical properties of plastics rather than on their significance.

In 1991 Carol Mancusi-Ungaro, then the chief conservator of the Menil Collection in Houston, USA, started the Artists Documentation Program to film interviews with artists in front of their works (Dupree, 2002). She asserted that artists and conservators shared respect for materials and the working properties of materials. The purpose of the free-form, wide-ranging interviews was not only to elicit technical information but also to document artists' reactions to their works as well as their feelings about the ageing of the work and about its future preservation.

The two symposia that first brought together conservators, conservation scientists, curators, art historians and artists to discuss the longevity of materials used in modern art and the future of art and art collection were 'Modern Art: Who Cares?' hosted by the Netherlands Institute for Cultural Heritage in 1997 (Hummelen and Sillé, 1999) and 'Mortality Immortality? The Legacy of 20th-Century Art' held at the Getty Center in Los Angeles in 1998 (Corzo, 1999). Prior to these conferences, it had been assumed widely by museum professionals that artists intended their works to last for many years whatever the materials used. These conferences highlighted the challenges associated with preserving short-lived plastics for longer than their designed durability. In the late 1990s various artists' viewpoints on useful lifetime and the conservation of deteriorated materials used in art were discussed and debated openly.

Collaboration with artists has both increased and complicated the options for conserving plastics in art. The multidisciplinary 'Foundation for the Conservation of Modern Art' was established in the Netherlands in 1996 to ascertain the information required to conserve modern art. A decision-making model was developed which weights the importance of various factors prior to proposing a conservation treatment (Bosma et al., 1999). Factors include the significance of the materials, the artist's intention, techniques used and the condition of the materials. In contrast to considering the condition of materials alone, as a conservator or conservation scientist may, the decision-making model relates any degradation of materials present to the change in significance of the artwork. It may be that a scratch or discolouration of plastics changes the significance of one artwork in which a high surface gloss is important, but has no influence on another artwork with a matte surface. Application of the decision-making model requires interviews with the artist or obtaining information about the artist's intention and opinions by other methods. The decision-making model has now been accepted as a significant tool in the conservation process of all forms of modern and contemporary art.

The conference 'Plastics – Looking at the Future and Learning from the Past' at the V&A Museum, London, in 2007 highlighted two issues concerning the acquisition of artworks and their conservation which have resulted directly from the interaction between museum professionals and artists. First, museums and art galleries are less willing than they were 10 years ago to acquire artworks containing plastics which are intended by the artist to last less than around 10–20 years or artworks which are likely to require extensive conservation resources (David, 2008). Such artworks can only be exhibited for short periods and are expensive to maintain. This policy is likely to result in the absence in the future, from museums and galleries, of works by particular artists or those constructed from particular plastics. Second, artists are willing to discuss the long-term stability of specific plastics and the options for their conservation with conservation professionals prior to selecting their working materials (Wells, 2008). Such collaborations are likely to result in artworks with longer useful lifetimes and fewer conservation needs than today, but may also limit the variety of future designs.

In addition to lack of resources, development of conservation techniques for plastics has been impeded by:

- the challenge of applying traditional, ethical principles of conservation to new, short-lived materials
- the idea that conservation of plastics should be approached differently to that of more traditional materials, and
- the need to investigate copyright and the moral rights of artists prior to conservation (Keneghan, 2005).

The conservation profession has been compared with the medical profession because both are practiced within a framework of ethical standards and principles to limit the extent of any damage. National and international organizations including ICOM and International Institute for Conservation (IIC) have produced guidelines with several common principles (Pye, 2001):

- any work undertaken should not affect the significance of the object – either the material structure or the intangible meaning
- the minimum amount of interventive work that will secure a satisfactory result should be undertaken on an object. If it is possible, any new materials introduced should be removable
- when any restoration is undertaken it should readily be detectable, even if not immediately apparent on superficial examination. All such work should be fully and clearly documented.

The major reason for underdevelopment of treatments for plastics which adhere to ethical guidelines is the high sensitivity of many plastics, especially when deteriorated, to organic liquids, aqueous solutions and water itself. It is unusual to observe the same high sensitivity in other organic materials such as paper, textiles and leather. In addition, coatings or adhesives which adhere successfully to plastics' surfaces either soften the substrate by dissolving or etching it to achieve bonding. Such treatments frequently change the appearance of the original, which may not be acceptable for cultural property, particularly works of art. Interventive conservation treatments applied to plastics have acquired a reputation as high risk and irreversible.

From a legal perspective, although a museum, gallery or private person has bought or otherwise acquired a contemporary artwork, the immaterial components still remain the property of the artist. Copyright and the moral rights of artists legally affect the type and extent of conservation of modern artworks. Copyright law protects all original works for 70 years after the death of the artist in European Union countries and 50 years in others. This suggests that most twentieth-century art is still protected by copyright and that the owners of such works are not free to treat the material as they wish (Dreier, 1999). In addition, the Berne Convention for the Protection of Literary and Artistic Works defines the moral right to the integrity of a work as 'the right … to object to any distortion, mutilation or other modification of, or other derogatory action … which would be prejudicial to the artist's honour or reputation' (Cornell Law, 2007).

Both copyright law and the Berne Convention must be considered prior to implementing a conservation strategy to counteract degradation of the artwork over time. If the artist agrees to the treatment proposed by the owner institute, there is no conflict. However, if the artist intended the work to be ephemeral from its conception, while the owner wishes to prolong its useful lifetime, the situation is more complex and not covered by established law.

1.4   Status of conservation of plastics

The conservation treatments for degrading plastics in use today are discussed in detail in Chapter 7. A brief overview of the status of conservation of plastics is presented here. The least stable plastics in museum collections have been identified and comprise cellulose nitrate and acetates, poly (vinyl chloride), polyurethanes and rubbers. Due to lack of resources, these 'high-risk' materials have been the focus of research projects looking at how to prolong the useful lifetime of plastics, at the exclusion of others. Polyester, polyethylene and polypropylene plastics are increasingly showing signs of degradation in museum collections and are expected to be the focus for conservation research projects in the near future (van Oosten et al., 2008). The main factors which initiate or accelerate degradation of all plastics have been identified as light, especially high-energy ultraviolet radiation, heat, oxygen and water.

Conservators consider two approaches to conservation when planning treatment to extend the useful lifetime of all materials, namely preventive and interventive. Preventive conservation may be defined as the mitigation of degradation and damage to cultural property through the formulation and implementation of policies and procedures for storage, exhibition, packing, transport, and use, integrated pest management, disaster management and reformatting or duplication (American Institute for Conservation of Artistic and Historic Works,

1997). Once initiated, degradation of plastics cannot be prevented, reversed or stopped, but only inhibited or slowed. For that reason, I prefer the term 'inhibitive' conservation to 'preventive'. Slowing the rate of degradation is realized by storing objects in an appropriate microclimate which removes the main factors causing degradation, particularly oxygen and moisture. Prolonging the useful lifetime of plastics by inhibitive conservation is possible today but is limited in scope and effectiveness. Appropriate microclimates have been identified for cellulose nitrate and acetates, poly (vinyl chloride), rubbers and polyurethanes (Shashoua, 2006). These are discussed in detail in Chapter 7. An inhibitive approach can help to prolong the useful lifetime of many objects at the same time. The majority of plastics objects in museums spend their useful lifetime in storage areas, so microclimates in stores can contribute greatly to the lifetime of plastics. Inhibitive conservation techniques are likely to comply with the ethical practices of professional conservators.

Interventive conservation treatments are those involving practical treatments applied as necessary to individual objects to limit further degradation and often to preserve their significance. Treatments include adhering broken sections, cleaning and strengthening surfaces. Although all surveys of the condition of objects containing plastics in the United Kingdom concluded that approximately 75 per cent of collections required cleaning, such conservation practices are poorly developed today (Shashoua and Ward, 1995). This area has been less developed than inhibitive treatments due to the high sensitivity of plastics to cleaning agents and techniques, solvents, adhesives and consolidants. The risk of causing irreversible damage to plastics is greater during interventive conservation than inhibitive. However, pressure from museum professionals who study and care for plastics, the commercial art market, private collectors and also from organizers of exhibitions is likely to accelerate progress in this field.

References

Albus, S., Bonten, C., Keßler, K., Rossi, G. and Wessel, T. (2007). *Plastic Art – A Precarious Success Story*. AXA Art Versicherung, Cologne.

Allen, N. S., Edge, M. and Horie, C. V. (eds) (1992). *Polymers in Conservation*. Royal Society for Chemistry, London.

American Institute for Conservation of Artistic and Historic Works (AIC) (1997). *Commentary 20: Preventive Conservation*. Available from: http://aic.stanford.edu/pubs/comment20.html [Accessed 2 April 2006].

Biefer, M. and Zgraggen, B. (1991). *Prophecies no.27*. Elwick Grover Aicken, Brighton.

Bosma, M., Hummelen, Y., Sillé, D., van der Vall, R. and Wegen, R. (1999). Decision-making model for the Conservation of Modern Art. In *Postprints of Modern Art: Who Cares?* Amsterdam (U. Hummelen and D. Sillé, eds) pp. 164–172. The Foundation for the Conservation of Modern Art and the Netherlands Institute for Cultural Heritage.

Bradley, S. M. (1994). Do objects have a finite lifetime?. In *Care of Collections* (S. Knell, ed.) p. 57, Routledge, London.

Brydson, J. A. (1999). *Plastics Materials*, 6th edition. Butterworth-Heinemann,Oxford.

Chiantore, O. and Rava, A. (2005). Conservare l'arte contemporanea – problemi, metodi, materiali, ricerche. Electa (http://www.electaweb.it), Italy.

Cornell Law (2007). *Berne Convention for the protection of literary and artistic works (Paris Text 1971), article 6bis.* [online]. Available from: http://www.law.cornell.edu/treaties/berne/6bis.html [Accessed 16 July 2007].

Corzo, M. A. (ed.) (1999). *Mortality Immortality? The Legacy of 20th-Century Art.* The Getty Conservation Institute, Los Angeles.

Dahlsen, J. (2007). *Recycled Bag Art* [online]. Available from: http://www.johndahlsen.com/thumbs/plastic_bags.html [Accessed 16 July 2007].

David, F. (2008). Challenge of Materials? A New Approach to Collecting Modern Materials at the Science Museum. In *Postprints of Plastics: Looking at the past & learning from the future.* London, 23–25 May 2007, publication expected in 2008.

Dreier, T. K. (1999). Copyright aspects of the preservation of nonpermanent works of modern art. In *Postprints of Mortality Immortality? The legacy of 20th-century art* Los Angeles 25–27 March 1998 (M. A. Corzo, ed.) pp. 63–66, Getty Conservation Institute, Los Angeles.

Dupree, C. (2002). *Impermanent Art* [online]. Available from: http://www.harvardmagazine.com/on-line/01027.html [Accessed 30 July 2007].

European Union Commission (2000). *Green paper on environmental issues of PVC (COM (2000) 469FINAL)* [online]. Available from: http://www.europa.eu.int/comm/environment/pvc/index.html [Accessed 24 February 2007].

Fakelite.com (2005). *Fakelite information for Bakelite collectors* [online]. Available from: http://www.fakelite.com/ [Accessed 20 August 2007].

Grattan, D. W. (ed.) (1993). *Saving the twentieth century: the conservation of modern materials.* Canadian Conservation Institute, Ottawa.

Hochfield, S. (2002). Sticks and Stones and Lemon Cough Drops. *ART news* [online] September 2002. Available from: http://artnews.com/issues/issue.asp?id=10108 [Accessed 30 July 2007].

Hummelen, Y. (1999). The Conservation of Contemporary Art: New Methods and Strategies? In *Postprints of Mortality Immortality? The legacy of 20th-century art* Los Angeles 25–27 March 1998 (M. A. Corzo, ed.) pp. 171–174, Getty Conservation Institute.

Hummelen, U. and Sillé, D. (eds). (1999). *Modern Art: Who Cares?* The Foundation for the Conservation of Modern Art and the Netherlands Institute for Cultural Heritage, Amsterdam.

ICOM-CC Modern Materials and Contemporary Art Working Group (2007). Available from: http://www.icom-cc.org [Accessed 6 July 2007].

Keneghan, B. (1996). Plastics? – Not in my collection. *V&A Conservation Journal* No. 21, October 1996, 4–6.

Keneghan, B. (2005). Plastics preservation at the V&A. *V&A Conservation Journal* [online] No. 50, Summer 2005. Available from: http://www.vam.ac.uk/res_cons/conservation/journal/number_50/plastics/index.html [Accessed 2 July 2007].

Koizumi, M. (2005). *PET project* [online] Available from: http://miwa.metm.org/PET_project/index.html [Accessed 16 July 2007].

Lavedrine, B., Gandolfo, J-P., Monod, S. and Grevet, S. (2003). *A Guide to the Preventive Conservation of Photograph Collections.* Oxford University Press, Oxford.

Learner, T. J. S. (2005). *Analysis of Modern Paints.* Getty Conservation Institute, Los Angeles.

Morgan, J. (1994). *A survey of plastics in historical collections.* Plastics Historical Society and The Conservation Unit of Museums and Galleries Commission [online]. Available from: http://www.plastiquarian.com/survey/survey.html [Accessed 7 March 2007].

Mossman, S. (1993). Plastics in the Science Museum, London: A curator's view. In *Postprints of Saving the Twentieth Century: The Conservation of Modern Materials Ottawa, 15–20 September 1991* (D. Grattan, ed.) pp. 25–35, Canadian Conservation Institute.

Nord, A. G., Lampel, K., Björling-Olausson, Jonsson, K., Franzon, M., Hallden-Tengner, C., Mattson, E. and Johansson, M. (2006). *Morgondagens kulturobjekt projekt foer bevarande av plastfoeremaal (Tomorrows objects – project to preserve plastics)*. Unpublished stage report for FoU-projekt. National Heritage Board of Sweden, Stockholm.

Plastics Historical Society (2007). *Plastiquarian* [online]. Available from: http://www.plastiquarian.com [Accessed 5 July 2007].

Pye, E. (2001). *Caring for the past.* James & James Ltd, London 2001.

Rossnes, G., Amundsen, K. and Hadland, G. (2002). *The Industrial Heritage Work in Norway* [online]. Available from: http://www.ihp.lt/gateway/no/no_index.html [Accessed 30 July 2007].

Quackenbos, H. M. (1954). Plasticizers in vinyl chloride resins. *Industrial and Engineering Chemistry*, **46** (6), 1335–1341.

Quye, A. (1993). Examining the plastic collections of the National Museums of Scotland. In *Preprints of Conservation Science in the UK* Glasgow, May 1993 (N. H. Tennant, ed.) p. 48, James & James Science Publishers, London.

Shashoua, Y. (2001). Inhibiting the deterioration of plasticised poly(vinyl chloride) – a museum perspective. Ph.D. thesis, Danish Polymer Centre, Technical University of Denmark.

Shashoua, Y. (2006). Plastics. In *Conservation Science Heritage Materials* (E. May and M. Jones, eds) pp. 185–210, RSC Publishing, Cambridge.

Shashoua, Y. and Ward, C. (1995). Plastics:modern resins with ageing problems. In *Preprints of Resins, Ancient and Modern* Aberdeen, September 1995 (M. M. Wright and J. H. Townsend, eds) pp. 33–37, SSCR, Edinburgh.

The Associated Press. (2007). *95 million pound sale of art gets European record for Sotheby's* [online]. Available from: http://www.iht.com/articles/ap/2007/02/06/arts/EU-A-E-ART-Britain-Art-Sale.php [Accessed 16 July 2007].

Then, E. and Oakley, V. (1993). A survey of plastic objects at the Victoria and Albert Museum. *V&A Conservation Journal*, **6**, 11–14.

Tímár-Balázsy, Á. and Eastop, D. (1998). *Chemical principles of textile conservation.* Butterworth-Heinemann, Oxford.

van Oosten, T. (1999). Here Today, Gone Tomorrow? Problems with plastics in contemporary art. In *Postprints of Modern Art: Who Cares?* Amsterdam (U. Hummelen and D. Sillé, eds) pp. 158–163, The Foundation for the Conservation of Modern Art and the Netherlands Institute for Cultural Heritage.

van Oosten, T., Shashoua, Y., Waentig, F. (eds). (2002). *Plastics in Art, History, Technology, Preservation.* Kölner Beiträge zur Restaurierung und Konservierung von Kunst- und Kulturgut Band 15. Siegl, Munich.

van Oosten, T., Bollard, C. and de Castro, C. (2008). Lights out!!! Research into the conservation of polypropylene. In *Postprints of Plastics: Looking at the past & learning from the future* London, 23–25 May 2007, publication expected in 2008.

Ward, P. (1997). *Fantastic Plastic – the kitsch collector's guide.* New Burlington Books, London.

Wells, P. (2008). Site-specific polythene:Experiments in Durability. In *Postprints of Plastics: Looking at the past & learning from the future* London, 23–25 May 2007, publication expected in 2008.

Wright, M. M. and Townsend, J. (eds). (1995). *Resins ancient and modern.* SSCR, Edinburgh.

Historical development of plastics

Summary

Chapter 2 describes the factors which were instrumental in the history of plastics and the development from natural materials to semi-synthetics and then to the synthetics we know today. Collections contain examples of all groups of plastics. They represent both the materials and technology available at different periods of the nineteenth, twentieth and twenty-first centuries. The future of plastics and, thereby, of future collections is presented.

The relevance of the historical development of plastics to conservation professionals is largely as a tool to date or help identify the types of materials or technologies represented in collections. Conservators and historians are also interested in the reasons and background to the development of plastics. A brief history of the development of plastics is presented in this chapter. More detailed information may be obtained from publications and the website of the Plastics Historical Society (Plastics Historical Society, 2006; Mossman, 1997).

The word polymer is derived from the Greek *poly* and *meros* meaning many and parts respectively. Scientists also use the word macromolecule meaning large molecule instead of polymer. The long chains of atoms which comprise polymers and the bonds within and between the chains determine plastics' properties. Polymers alone are not plastics. Plastics are polymers or macromolecules which have been modified with additives and shaped to convert them from liquids to solids which adopt a dimensionally stable form.

Although we consider plastics to be new materials, only semi-synthetic and synthetic plastics are modern (Table 2.1). Naturally occurring polymers, those formed by plants, trees, insects and animals, were first documented as coatings and waterproofing agents in the form of bitumen, referred to as slime in the Old

Table 2.1 Chronological development of commercial polymers		
Date	**Material**	**Major applications**
Pre-1800	natural polymers including cellulose, wool, natural rubber, gutta percha, shellac	clothing, jewellery, electrical insulation, waterproofing, coatings
1839	vulcanization of natural rubber	waterproof coatings, clothing, balloons, toys
1846	cellulose nitrate	explosives, e.g. 'gun cotton'
1851	ebonite or hard rubber	Jewellery, knife handles
1870	celluloid (plasticized cellulose nitrate)	fake ivory and tortoiseshell, combs, jewellery, collars
1889	cellulose nitrate photographic film base	photograph negatives, movie film base
1899	casein-formaldehyde	buttons, cutlery handles
1907	phenol-formaldehyde (Bakelite)	electrical insulation, radio cabinets, telephones, cookware, laminates
1908	cellulose acetate photographic film base	safety movie film base
1912	regenerated cellulose sheet (cellophane)	packaging film
1927	poly (vinyl chloride)	electrical insulation, waterproof clothing, flooring and wallpaper, packaging, vinyl records
1929	urea-formaldehyde	radio casings, insulation, kitchenware
1931	poly (methyl methacrylate)	glazing, protective screens, advertising signs
1937	polystyrene	thermal and electrical insulation, contact lenses, advertising signs
1938	nylon-66 fibres	nylon stockings, parachutes, electrical insulation
1939	melamine-formaldehyde	kitchenware, laminates
1941	low density polyethylene	electrical insulation, packaging, food use
1942	unsaturated polyesters	drinks containers, food packaging, composites, electrical insulation, boats, sports and medical equipment

Table 2.1 Continued		
Date	**Material**	**Major applications**
1943	fluorocarbons (Teflon)	electrical and heat insulation, cookware, protective clothing
	silicones	thermal and electrical insulation
	polyurethanes	electrical and thermal insulation, protective clothing and packaging, rubbers
1947	epoxy	composites, electrical insulation
1948	acrylonitrile-butadiene-styrene copolymer	glazing, packaging, advertising signs
1957	high-density polyethylene	electrical insulation, packaging, food use
	polypropylene	electrical insulation, food packaging furniture, car bumpers
	polycarbonate	glazing, CDs, DVDs
1964	poly (phenylene oxide)	microwave insulation components, water pumps, hot water tanks
1970	poly (butylene terephthalate)	gear wheels, pump housings, textile bobbins
1971	poly (phenylene sulphide)	sterilizable medical equipment, hair dryers

Testament (Brydson, 1999). Bitumen was also used in the preparation of mummies in Ancient Egypt. Mummies were wrapped in cloth dipped in a form of bitumen prior to being hardened by exposure to sunlight (Couzens and Yarsley, 1956). Shellac from insects was used to prepare cast mouldings by the ancient Indians. Sealing wax based on shellac was used in Europe from the Middle Ages. Gutta percha, obtained as latex from the bark and leaves of *Pal. oblongifolium* trees, achieved great importance as an underwater cable insulation material in the middle of the nineteenth century and was replaced by synthetic polymers as recently as 1940.

In 1839 and 1840, Charles Goodyear and Thomas Hancock discovered that heating natural rubber with a few per cent of sulphur allowed it to remain both elastic over a wide range of temperatures and resistant to solvents, while adding high quantities of sulphur produced hard ebonite or vulcanite. This discovery was a milestone in the history of plastics. Ebonite was the first plastics material to be produced by chemical modification of a natural polymer and was used to form buttons, jewellery, and match boxes. The theory of the crosslinking reaction behind the production of ebonite was not understood at that point.

Natural polymers were less dense, more robust and cheaper than glass and ceramics but were not available in the wide range of colours and forms required by artists. Such needs inspired the development of synthetic alternatives. The other influence on synthetic polymers were the developments in practical synthetic organic chemistry in the first half of the nineteenth century. The first synthesis of an organic chemical was carried out by the German chemist, Wöhler, in 1824.

2.1   Development of semi-synthetic plastics

Semi-synthetic plastics are natural polymers which have been treated chemically to modify their properties with the aim of producing physically stable, mouldable products. The Swiss chemist, Schönbein, unwittingly prepared the first semi-synthetic polymer by treating paper (cellulose) with a mixture of nitric and sulphuric acids in 1846. The resulting cellulose nitrate was soluble in organic liquids and was softened by heat. It was highly flammable and only thought to be useful as an explosive, which was named 'gun cotton'.

An English metallurgist, Alexander Parkes, applied his practical experience of materials to the challenge of moulding cellulose nitrate (CN). Parkes discovered that moulded forms shrank unacceptably on drying but the addition of heated camphor to a solution of CN solved this problem by acting as a plasticizer or softener. Camphor is a white crystalline solid which can either be extracted from the trunks, roots and branches of *cinnamomum camphora* trees in Asia and Florida, or can be synthesized. Camphor sublimes (evaporates from the solid state) at room temperature and is known for its musty smell which deters moths and animals. CN was launched as buttons, combs and jewellery under the tradename Parkesine at The International Exhibition in London in 1862. Unfortunately, Parkes marketed cellulose nitrate as a substitute for rubber, ebonite and gutta percha which were already limited markets and his business collapsed after 2 years.

Around 1870, John Wesley Hyatt, an American printer, attempted to win a $10 000 prize by developing a substitute material for ivory billiard balls. Hyatt mixed cellulose nitrate with camphor, heated it under pressure and shaped it. The product, known as celluloid, could be used to form boxes, wipe-clean linen shirt collars and cuffs, ping-pong balls, dolls and dental plates, which were made from ebonite in the early nineteenth century (Figure 2.1). Because cellulose nitrate softens on warming (thermoplasticity) it was not an ideal material for false teeth, which curled when the wearer drank hot coffee. Its

Figure 2.1
Dolls were made
from cellulose nitrate
between 1900 and the
mid 1930s.

Figure 2.2
Cellulose nitrate was
originally developed
to imitate expensive
natural materials
such as tortoiseshell.
Decorative hair combs
made from celluloid in
the 1920s look similar to
genuine tortoiseshell.

strength as a material was its ability to mimic the appearances of natural materials including ivory, marble, mother-of-pearl and tortoiseshell. In contrast to Parkes, Hyatt concentrated on producing exclusive rather than inexpensive products. However, celluloid was still used only to replace natural materials rather than to develop new products (Figure 2.2).

The most important commercial application for celluloid was as a film base due to its high transparency, homogeneity and flexibility. The film company,

Eastman Kodak, started to develop celluloid film in 1889 and by 1920 the movie industry was dependant on the new material. However, cellulose nitrate's high flammability and ability to soften on heating were in conflict with its use as a film base and alternatives were sought. Today, CN is mainly used as an adhesive, as inexpensive automotive retouching paints and nail lacquer, due to its high solubility in acetone.

The development of cellulose acetate solved the flammability problem in film. In 1869, the German chemist, Schutzenberger, acetylated cellulose by treating it with acetic acid instead of the nitric acid used to prepare cellulose nitrate, but the reaction was not scaled up until 1905. Its early uses included non-flammable or 'safety' film bases and dope to stiffen and waterproof the fabric wings

Figure 2.3
Cellulose acetate was used to imitate natural, expensive materials such as mother-of-pearl as shown by this spoon handle made in the 1940s.

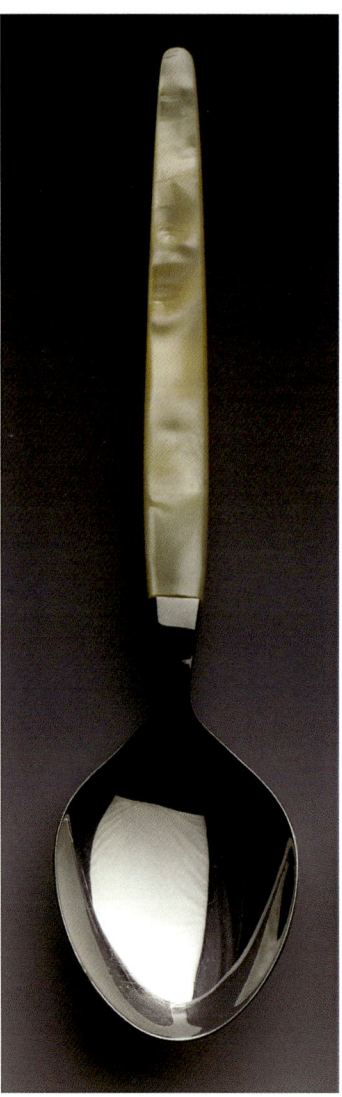

and fuselage of early aeroplanes. Cellulose acetate, like cellulose nitrate, was used to imitate expensive, natural materials (Figure 2.3). It was initially fabricated like celluloid in the form of rod, sheet or tube but in 1919 became available as a moulding powder exhibiting various degrees of hardness which could be quickly and economically shaped by injection moulding (Kaufmann, 1963). Cellulose acetate was produced as film and three dimensional products under the tradenames Bexoid, Clarifoil and Tenite. Cellulose acetate was formed into Rayon® fibres for synthetic textiles at the start of the 1920s. Cellulose acetate is still used in film base and synthetic textiles today.

Another semi-synthetic substitute for cellulose nitrate was casein-formaldehyde. In 1899, Krische and Spitteler developed a whiteboard for German schools by coating paper with casein, a protein found in cow's milk. Treating casein with formaldehyde hardened it and imparted resistance to moisture. Casein was extruded into rods, rolled into sheets and coloured prior to treatment with formaldehyde. The brilliant colours and patterns available made casein a suitable material for products such as buttons, buckles, fountain pen barrels and knitting needles with tradenames Syrolit, Lactoid, Erinoid, Aladdinite, Amaroid and Galalith (Figure 2.4). Casein's disadvantages as a material were its sensitivity to water and heat. Because milk is one of the raw materials, production of casein-formaldehyde has long been associated with the dairy industry. At the start of World War II supplies of casein granules could not be obtained from Europe and were required for buttons for military uniforms. An alternative source of casein was found in Argentina. Production of casein-formaldehyde ceased in the United Kingdom in around 1980 and very little is manufactured in the USA.

Figure 2.4
Casein was used widely to make buttons. They were coloured and polished in sought-after designs.

2.2   Development of synthetic plastics

The polymers synthesized before the start of the twentieth century were all based on natural polymers and most had been discovered or developed into commercial materials by trial and error. Scientists and others developing polymers had shown a talent for making empirical discoveries before the supporting theories were understood. Elucidating the theory of organic chemical reactions in the 1930s provided tools to predict the outcome of reactions and the development of fully synthetic polymers became more systematic. The leading organic chemists of the nineteenth century, such as Bayer, Michael and Kleeberg, recognized that phenol would react with formaldehyde and produce water as a by-product, but were not able to calculate how much of each starting material was necessary to produce a crosslinked product rather than a soluble coating.

Baekeland, a chemist born in Belgium and living in the USA, discovered that heating a mixture of phenol and formaldehyde together with a catalyst while applying pressure produced a hard, insoluble, non-melting (thermosetting) material. Application of pressure prevented the water and steam by-products from creating bubbles in the polymer. The first fully synthetic polymer, phenol-formaldehyde, was called Bakelite and a patent was taken out in 1907. When the patent expired in 1927, phenol-formaldehydes with names such as Nestorite, Mouldrite, Isolit and Nokait were produced. Bakelite could be worked, was resistant to acids and organic liquids, resisted heat and was an excellent electrical insulator. It was used to make bowling balls, records, radio housings, telephone housings, electrical insulation and cookware (Figure 2.5). Bakelite was also used as a binder in composites together with textiles and paper. Products included Formica laminates. Although there is no evidence that Baekeland appreciated what polymers were, he seems to have understood the principle of functionality, i.e. the idea that varying the ratio of phenol to formaldehyde affects the properties of the final material.

Figure 2.5

Bakelite was the first completely synthetic plastic. This Bakelite speaker from the 1930s was made by the electrical company Laurits Knudsen.

In Britain, similar experiments were being carried out by a British inventor, Sir James Swinburne, whose search for a material with good electrical properties led him to develop similar materials to Bakelite. His experiments were less complete than those of Baekeland but the two collaborated in the 1920s to develop the Bakelite business in Britain. The success of phenol-formaldehyde mouldings stimulated research into other polymers.

The dark colours of phenolic resins meant that plastics were only available in black, red and shades of brown. The search for a colourless resin with similar properties to phenolics led to the development in the 1920s and 1930s of urea- and thiourea-formaldehydes (Figure 2.6). In 1918, Hans John reacted urea with formaldehyde. The reaction was further studied by Pollak and Ripper between 1920 and 1924 in an unsuccessful attempt to produce an organic glass. Around the same period, the British Cyanides Company (later known as British Industrial Plastics) investigated the condensation of thiourea with formaldehyde. When combined with cellulose fillers and suitable colourants, urea- and thiurea-formaldehydes made possible the production of articles such as trays, cups, picnic-ware and lampshades in white and brilliant colours for the first time. These products were made under tradenames including Beatl, Bandalasta, Scarab, Mouldrite U and Lastalonga. In common with the phenolic resins, urea-formaldehyde also found important industrial applications in varnishes, laminates and adhesives.

Figure 2.6
Bakelite was only available in dark colours (left of picture). In the 1930s, urea-formaldehyde was developed and, for the first time, pale-coloured plastics became available (radio to right of picture). The dials of both radios are made of cellulose nitrate.

The development of melamine-formaldehydes in the mid 1930s completed the family of thermosetting formaldehyde condensation polymers. The melamines closely resembled the urea- formaldehyde plastics in their general properties and colour range, but they showed greater resistance to heat, water and detergents. The porcelain-like appearance of mouldings made it particularly attractive for moulding cups, saucers, plates and similar domestic items although they were more costly than their equivalents in urea-formaldehyde.

Poly (vinyl chloride), better known as PVC, was first polymerized by Baumann in 1872 but was only of academic interest at that time and was not associated with any practical applications. Russian chemist, Ostromislensky, patented the polymerization of PVC in 1912. The polymer was considered a possible replacement for cellulose nitrate because it was less flammable, so patents proposed its application to film, buttons and combs (Kaufmann, 1963). However, the first attempts at industrial scale production were unsuccessful. PVC's tendency to decompose at processing temperatures between 160°C and 180°C delayed its further development. One solution to this difficulty was to copolymerize vinyl chloride with vinyl acetate and a patent was obtained in 1928. Another notable solution was the incorporation of plasticizers or softeners in PVC which lowered its softening temperature. This allowed milder processing temperatures and increased flexibility of the finished products. Plasticizers, heat and light stabilizers, and other additives have contributed greatly to the continued success of PVC because a stable, long-lasting plastic based on this polymer cannot be produced without them.

B. F. Goodrich developed techniques to make sheets and adhesives from PVC in 1926. Pilot-scale production started in 1937 in Germany and full industrial-scale production was in progress one year later. PVC production in the USA followed shortly after. The number of applications for rigid and plasticized PVC developed rapidly after 1940. Despite its higher cost, PVC replaced natural rubber as electrical cable insulation due to its easy production methods, non-flammability, resistance to oils and water and because, unlike rubber, it didn't corrode copper wire (Kaufmann, 1963). PVC flooring was used in Germany because of its high resistance to scuffing. PVC's ability to withstand attack by acids was utilized in the chemical industry prior to World War II. After the war, there was an excess of PVC, which was diverted into packaging, clothes and shoes (Figure 2.7). Today PVC, in both rigid and plasticized forms, comprises one of the largest tonnage plastics materials, despite doubts about its stability and concerns since the 1980s about the potential of its plasticizers to damage children's health and its manufacturing processes to damage the environment.

Like PVC, polystyrene also experienced a long delay between synthesis and commercial development. Although polystyrene is thought to have been synthesized first by Simon in 1839 who thought it was an oxidation product of styrene, Kaufman suggests that its discovery was earlier (Kaufman, 1963). Bonastre described polystyrene in a publication in 1831 although it is unknown if he was the discoverer. There were no commercial applications for polystyrene in the 1800s. Commercial production was delayed by the need to prevent polymerization of the monomer styrene during storage. Commercial interest in polystyrene started in 1911 in England where it was considered as a substitute for celluloid, without success. The development of injection moulding opened many new markets for polystyrene in the 1930s, particularly in Germany. Polystyrene was copolymerized with butadiene to make synthetic rubber long before World War II. Polystyrene is a glass-clear plastic, a good electrical insulator, resistant to acids and alkalis, a barrier to water and has high tensile strength even at low temperatures. It is used to make advertising signs, contact lenses, fridge linings and insulating foams (Figure 2.8).

Polyethylene was the unexpected product of an organized research project. In 1933, the effect of high pressure on chemical reactions was studied by

Figure 2.7
This jacket from the 1960s looks like leather but is made from plasticized PVC. 'Leatherette' was considerably cheaper than leather and available in bright colours.

Figure 2.8
This coffee percolator (6 cm high for a doll's house) is made of polystyrene. The coffee jug demonstrates the glass-like appearance of polystyrene.

chemists at the Alkali Division of Imperial Chemical Industries (ICI). Fawcett and Gibson, the chemists involved in the research project, noticed the formation of a white, waxy solid when ethylene gas was used in experiments. The solid was identified as polyethylene. In trying to reproduce the polymerization, it was discovered that oxygen was necessary for the reaction to take place and had only been present due to a leak in the original apparatus. Polyethylene is an excellent electrical insulator and has good chemical resistance. Its value as an underwater cable insulator was quickly recognized and industrial production started in September 1939, just prior to the outbreak of World War II. Today, polyethylene is the largest tonnage plastic and Tupperware food containers are one of its best known applications (Figure 2.9).

Figure 2.9
Tupperware food containers have been made of polyethylene since the 1950s.

The other development by ICI laboratories which made a vital wartime contribution was poly (methyl methacrylate), widely known as acrylic, Perspex or Plexiglas. First produced commercially in the UK in 1934, its rigidity, transparency and shatter-resistant properties were soon in high demand for glazing, aircraft canopies and protective screens because acrylics provided a lower density, weather-resistant alternative to glass (Figure 2.10). Production of poly (methyl methacrylate) increased 1200 per cent during the period 1940–43 due to the escalation in production of aircraft (Kline, 1944).

World War II increased both the demand and the number of applications for plastics. Just prior to the war, plastics were not only developed as substitutes for materials which were in short supply, notably rubber and gutta percha, but also to out-perform the then available plastics. Wartime demands encouraged scientists to develop polymers that could meet the demands of the military, electronics and food industries. Before 1940, plastics had mainly been used to make decorative and ephemeral pieces, but the war demanded more performance-based

Figure 2.10
Poly (methyl methacrylate) is used widely in glazing, screens and signs due to its glass-like optical properties as shown by this advertising silhouette of the Eiffel Tower.

applications including gas masks, protective helmets, electrical insulation and radio components. In addition to new materials, a deeper understanding of polymer chemistry was developed (Brydson, 1999). Polymers developed after the 1940s were based on scientific principles rather than empiricism.

In the USA in the 1930s, an intensive research programme resulted in the development of new polymers, including nylon used to make clothing and parachutes, and a non-stick polymer with high temperature resistance in the form of polytetrafluoroethylene, widely known by its trade name Teflon. Polytetrafluoroethylene was initially used as insulation in radios and electrical cables. Other developments of the 1935–1945 period included silicones, widely used as water repellents and in heat-resistant paints, and epoxies, which have outstanding properties of adhesion and chemical resistance (Figure 2.11). Polyesters were developed during World War II. They had been discovered around 1847 by the Swedish chemist, Berzelius, but were first produced industrially in 1941. In 1942 the first laminating polyester was used to make insulators for radar antennae. Glass fibre-reinforced polyester was used in the war to construct boats and later used for design furniture (Figure 2.12).

Figure 2.11
Silicone rubbers are flexible, water and chemical resistant, and are used to make the campaign wristband bracelets which have become fashionable in the twenty-first century. The wristband shown was sold to raise money for research into cancer and bears the message 'livestrong'.

Figure 2.12
A model of the Panton chair, designed in 1959 and 1960 by Verner Panton, and constructed in glass-reinforced polyester. The Panton chair was the first in design history to have no back legs and was moulded in one piece. The model is one sixth full size (14 × 8.3 × 6.6 cm).

The car industry also began to employ the new materials and a study by the large polymer company DuPont shows that more than 120 different car parts were constructed in plastic by 1942 (Fleck, 1952). Although plastic car bodies are thought to be products of the twenty-first century, Ford had already made a prototype of a car body in glass-reinforced polyester by 1941 (Hansen and Serin, 1989).

After World War II the newer polymers such as poly (vinyl chloride), poly-ethylene and polystyrene could be produced in large quantities at low cost. They replaced the older polymers such as cellulose nitrate as well as some of the traditional construction materials in some applications including metals, wood, leather and glass to give both technical and economic benefits. Plentiful and inexpensive, plastics lost their previous reputation as special and valuable materials and gained that of worthless and everyday.

Few new polymers were introduced in the period 1945–55 but they included high density polyethylene with superior properties to low density polyethylene and polypropylene. Both were the products of collaboration between research institutions and the plastics industry. Polycarbonate was first developed in 1953 and commercially produced in 1959 under the name Macrolon (Ekström, 1985). Polycarbonate is impact resistant and transparent, has good electrical insulation properties and can withstand temperatures up to 150°C. It has been employed widely in the electrical and electronics industry and had largely replaced aluminium in household appliances by the end of the 1970s. The most widely recognized applications for polycarbonate are as the base materials for CDs and DVDs. New plastics developed since 1970 have been highly specialized and applied to high temperature engineering and medicine rather than for general applications.

Another change in the plastics industry over time has been the sourcing of raw materials. Prior to World War II, cellulose nitrate and acetate were pro-duced from the cellulose in vegetable matter. After the war, ground nuts and vegetable oils were used to produce acids for manufacture of nylon in the USA. Polyethylene was obtained from sugar cane, ethanol and ethylene. Until the mid 1950s, the main raw material source for the European plastics industry was destructive distillation of coal, primarily from Germany. By the outbreak of World War II, the petrochemical industry was supplying some monomers including styrene and butadiene and the number increased yearly until the 1970s. The oil crisis in 1973 and the acceptance that sources of petroleum are not inexhaustible have motivated research into alternative sources for raw materials for plastics.

2.3 Tomorrow's plastics

Plastics have become indispensable to modern living and no longer represent only luxury and novelty as they did in the nineteenth century. Almost 35 million tonnes of synthetic polymers are produced annually in the USA and growth is expected as long as petroleum and other feedstocks last and consumer demand continues. In 1999, the USA accounted for 37 per cent of total plastics pro-duction, Western Europe 33 per cent and Japan 15 per cent (Brydson, 1999).

Modern society is highly dependent on petroleum as a cheap source of energy and raw materials. Price increases attributed to shortages of petroleum and in the absence of a viable alternative are likely to affect the plastics industry. Although plastics can be made from renewable vegetable products including seaweed, oat husks, soya bean and molasses, such raw materials are not as convenient as petroleum and further research is required to improve their effectiveness.

The effectiveness and safety of additives used to impart polymers with physical and chemical properties have been investigated since man-made plastics were first synthesized and are likely to remain areas of development in the future. Since the 1980s, health concerns about the plasticizers present in plastic packaging for food have raised consumer awareness and resulted in the enforcement of legal rulings. In Europe, phthalate plasticizers are no longer permitted in food packaging or in toys and other articles designed for use by young children, due to the association of phthalates with hormonal disruptions in the body that could result in infertility, asthma and allergies (European Parliament Council, 2005). As a result, technically and economically suitable alternatives to phthalates are undergoing evaluation. An example is a plasticizer synthesized from hardened castor oil and acetic acid (ElAmin, 2006). Manufacturer Danisco claim Grinsted Soft-N-Safe® to be colourless, odourless and completely biodegradable and can be used for food and beverage packaging without the need to modify equipment used to process PVC with phthalates.

Composites rather than the development of new polymers seem likely to play an important role in future materials. Both thermoplastic and thermosetting plastics reinforced with glass, carbon and nylon fibres have already made their mark on products from racing cars to tennis rackets. Future building materials which have been on trial extensively in Germany and Eastern Europe include a composite of polystyrene and concrete. Polystyrene foam blocks can be used to construct houses (Jensen, 2007). When the blocks are in the desired arrangement, concrete is poured into the hollows to form a construction with a similar tensile strength to brick. The resulting 'magu' buildings are rapidly constructed, low cost and energy efficient due both to the good insulating properties of polystyrene and the mechanical properties of concrete.

The car of the future could well have a body made from reinforced plastics, springs made from glass-reinforced plastics and plastics components in the engine. Though it may never be possible to realize a practical all-plastics car, the world's manufacturers are increasingly turning their attention to developing new mass production techniques in plastics. Composites are increasingly replacing metal components in processes from food production to nuclear reprocessing, household appliances and electronics, and human tissue-compatible materials.

Modifying the surfaces of plastics to impart specialized properties is also likely to be a growing area. Acrylonitrile butadiene styrene (ABS) has recently been surface treated with silver compounds to impart anti-bacterial properties which remain active for the lifetime of the plastic (Shaozao, 2005). The treated plastic has been used to make food containers which can keep their contents fresh for days.

Plastic electronics comprise devices made from conductive plastics instead of silicon which requires high temperatures and expensive clean room facilities. Highly conductive plastics were first discovered in the 1960s. Circuits can be printed onto polyester sheets to act as the control circuits for large flexible electronic paper displays such as daily newspapers. Polyester newspapers can fit into a jacket pocket, do not get wet in the rain and can be readily recycled. Such devices are currently under development, can store the data equivalent of thousands of books and are likely to be widely available by 2010. Currently available plastic electronics do not work sufficiently fast to replace silicon in microchips, but improving their performance is a new area of research (BBC, 2007).

A challenge which has faced the plastics industry since the 1970s, and is likely to continue, is concern for the environment. The increasing awareness of the need to conserve resources and to reduce pollution has led to criticism of the practices of the plastics industry. In addition, the lifecycle of plastics is of concern. A report from a United Nations environmental programme published in 2005 stated that there were an average of 46000 pieces of plastic at, or close to, surfaces of every square mile of ocean (Global Marine Litter Information Gateway, 2005). Large plastic items such as drinks bottles and packaging have well-known effects on sea life, strangling birds and fish, while smaller particles of plastics are thought to absorb toxic chemicals from seawater, poisoning the creatures that swallow them.

Analysing samples of sand from 20 beaches in the United Kingdom has revealed acrylics and polyesters which came originally from deep water (McKee, 2004). The number of small plastic particles on beaches is thought to have tripled since the 1960s and 70s.

As a result, the development of reusable materials and techniques to recycle plastics has attracted attention since the 1980s and is likely to affect the future demands on plastics. The applications for recycled plastics are expanding with recycled polyethylene competing with wood for garden furniture, decking and railway sleepers. Polyethylene requires little maintenance, unlike wood, and is waterproof, resistant to graffiti and cheaper than wood. Plastic railway sleepers constructed from foamed polyurethane reinforced with glass fibres are used to reduce noise from trains in Japan (PlasticsResource.com, 2006).

Reusable plastic sheets are being used in Japan to replace paper in printers thus reducing carbon emissions. Heat sensitive pigments are applied to polyester (poly (ethylene terephthalate)) sheets. By exposing the pigments to various temperatures, it is possible to apply and erase black and white text and images with a resolution of 12 dots per millimetre or 300 dots per inch (BBC, 2006). Thermal printing technology was first developed in the 1970s for fax machines. Each polyester sheet can be used up to 500 times.

Australia has used polypropylene currency notes since 1988 and, by the end of 2004, twenty-two countries had followed suit (Museum of Australian Currency Notes Virtual Tour, 2006). Polypropylene bank notes show advantages in security, durability and recycling over paper. Coloured background designs are offset printed simultaneously on both sides of the notes using specially developed inks (Figure 2.13). Additional designs such as portraits are intaglio printed using engraved metal plates on both sides of the notes, one at a time, producing raised print. Serial numbers are added using the letterpress printing process prior to applying two coats of varnish and cutting notes to size.

Figure 2.13

More than 22 countries use polypropylene bank notes for their increased security and durability compared with paper. The five-dollar note for Brunei Darussalem was introduced in 1996.

In conclusion, it is likely that more composite materials, recycled plastics and formulations containing fewer environment-and health damaging additives will be present in collections of the future. Plastic electronics are likely to raise the technological complexity of plastics materials in collections.

References

BBC (2006). *Plastic paper to 'cut' emissions* [online]. Available from: http://news.bbc.co.uk/1/hi/technology/6174052.stm [Accessed 8 January 2007].

BBC (2007). *Q&A: Plastic electronics* [online]. Available from: http://news.bbc.co.uk/1/hi/technology/6227455.stm [Accessed 8 January 2007].

Brydson, J. A. (1999). *Plastics Materials*, 6th edition. Butterworth-Heinemann, Oxford.

Couzens, E. G. and Yarsley, V. W. (1956). *Plastics in the Service of Man*. Pelican, Suffolk.

Ekström, G. (1985). Ett glasklart alternativ-Polykarbonat (A glass-clear alternative-polycarbonate). *Plastforum Scandanavia*, **6**, 94.

ElAmin, A. (2006). *Biodegradable plasticiser developed as phthalate replacement* [online]. Available from: http://www.packwire.com/news/printNewsBis.asp?id=64903 [Accessed 7 March 2007].

European Parliament Council. (2005). *Directive 2005/84/EC* [online]. Available from: http://eur-lex.europa.eu/LexUriServ/LexUriServ.do?uri=CELEX:32005L0084:EN:NOT [Accessed 24 February 2007]

Fleck, H. R. (1952). *The Story of Plastics*. Burke Publishing Co. Ltd., New York.

Hansen, P. A. and Serin, G. (1989). *Plast-fra galanterivarer til 'high-tech'. Om innovationsudviklingen i plastindustrien. (Plastic – from fine wares to high-tech innovations in the plastics industry)*. Special-Trykkeriet Viborg A/S, Denmark.

Global Marine Litter Information Gateway. (2005). *Marine litter* [online]. Available on: http://www.unep.org/regionalseas/marinelitter/ [Accessed 2 September 2007].

Jensen, M. J. (2007). Murstenen er lavet af flamingo (Bricks made from polystyrene foam, in Danish). *MetroXpress newspaper*, 20 February 2007, p. 22.

Kaufmann, M. (1963). *The First Century of Plastics-celluloid and its sequel.* Chameleon Press Ltd., London.

Kline, G. (1944). The Chemists' Wonderland – Plastics Through the Looking Glass. *Chemical and Engineering News*, **11**, 891.

McKee, M. (2004). Rising tide of micro-plastics plaguing the seas. New *Scientist*, 6 May 2004 [online]. Available from: http://www.newscientist.com/article/dn4966-rising-tide-of-microplastics-plaguing-the-seas.html [Accessed 7 March 2007].

Mossman, S. T. I. (1997). *Early plastics: perspectives 1850–1950.* Cassell, London.

Museum of Australia. (2006). *Currency Notes Virtual Tour: Polymer note exports* [online]. Available from: http://www.rba.gov.au/Museum/Displays/1988_onwards_polymer_currency_notes/export_of_notes.html [Accessed 7 March 2007].

PlasticsResource.com. (2006). *Recycled plastic lumbar* [online]. Available from: http://www.plasticsresource.com/s_plasticsresource/sec.asp [Accessed 7 March 2007].

Plastics Historical Society. (2006). *Plastiquarian* [online]. Available from: http://www.plastiquarian.com [Accessed 8 March 2007].

Shaozao, T. (2005). *Gongcheng Suliao Yingyong* (Journal written in Chinese with English abstract), **33** (5), 13–17.

Technology of plastics production

Summary

Chapter 3 describes the various components of plastics and how they are transformed into three-dimensional functional solids. The chapter is not intended as a technical manual from which polymers can be synthesized and plastics prepared but as an overview of selected procedures and materials relevant to museum professionals concerned with plastics. Knowledge of plastics technology helps to date plastics objects, to interpret chemical analysis of plastics and to understand the causes of deterioration. Polymerization, additives and shaping processes contribute to the physical and chemical properties, useful lifetime and degradation pathways of plastics. The optical, chemical and physical properties of plastics are detailed in Chapter 4.

Polymers are very high molecular weight materials formed by repeatedly joining low molecular weight units. Polymers alone are not plastics. Plastics are polymers which have been modified with additives and shaped to convert them from liquids to solids adopting a fixed form. Polymerization reactions, additives and shaping processes contribute to the physical and chemical properties, useful lifetime and degradation pathways of plastics. Preparation of semi-synthetic and synthetic polymers, types and properties of additives and the shaping processes for three-dimensional plastics are described in this chapter, while their properties are discussed in Chapter 4.

The nomenclature used by polymer scientists is based on the common name of the reactant monomer preceded by the prefix 'poly'. For example, polystyrene is the most frequently used name for the polymer derived from the monomer commonly known as styrene, and polyethylene is derived from the monomer ethylene. In the 1970s and 80s, the international body responsible for the systematic nomenclature of chemicals, International Union of Pure and Applied Chemistry (IUPAC), developed a system which names the components of the repeat unit, rather than the monomer, and arranges them in a prescribed order. Following the IUPAC system, polystyrene becomes poly (1-phenylethylene). The IUPAC system of nomenclature is not universally used, particularly by industry. Monomer-based nomenclature will be used here for ease of comparison with other publications. The traditional naming of polymer 'families' by the type of linkage connecting the polymers will also be adopted here. For example polycarbonates are polymers containing the carbonate linkage.

3.1   Preparation of polymers

The first man-made polymers were developed between the end of the nineteenth and the start of the twentieth centuries. They were known as semi-synthetics and formed the technological bridge between natural and fully-synthetic polymers. Semi-synthetic plastics were made by chemically treating a natural polymer to modify its properties, with the aim of producing a physically stable, mouldable product. In 1909, the first fully-synthetic polymer was produced commercially by reacting two chemicals together, namely phenol and formaldehyde, neither of which are natural polymers, to form the plastic known widely as Bakelite. Today, 50 different synthetic polymer types are included in around 60 000 plastics formulations with those based on polyethylene and poly (vinyl chloride) PVC competing for highest worldwide consumption (Quye and Williamson, 1999).

3.1.1   Preparation of semi-synthetic polymers

Cellulose-based plastics, particularly cellulose nitrate and acetates, were the most commercially important semi-synthetics up to the 1940s and were used as the base for photographic film, textile fibres, moulded goods and in lacquers. Naturally occurring polymer cellulose in the form of cotton linters or wood pulp is chemically treated to increase its solubility. Cellulose has a high molecular weight of between 100 000 and 500 000 and an empirical formula $C_6H_{10}O_5$. Casein-formaldehyde is the only protein-based moulded plastic that achieved commercial success. It is based on cow's milk and is still produced in very small quantities for specialist items such as hand-coloured buttons.

Cellulose nitrate (CN)
Schonbein showed in 1848 that cellulose treated with nitric acid in the presence of sulphuric acid was converted to nitrate esters. These are sometimes called nitrocellulose. This is a misnomer because they contain C—O—N bonds rather than the C—N bonds found in true nitrocompounds. During preparation

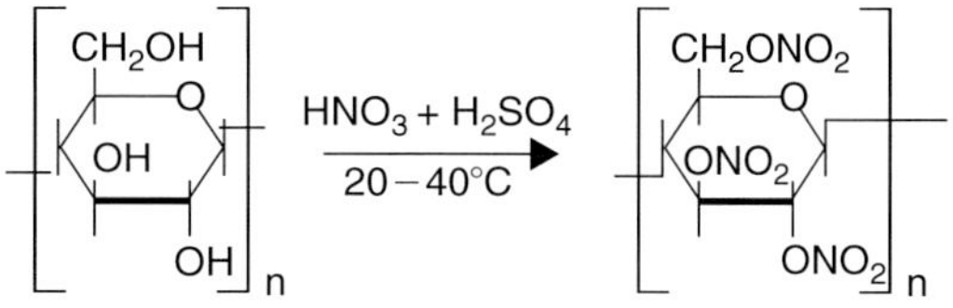

Figure 3.1

Preparation of cellulose nitrate involves replacing the hydroxyl (OH) groups on each of the many repeat units of cellulose. The number of repeat units (n) lies between 2000 and 26 000.

of cellulose plastics, the hydroxyl groups (OH) on the cellulose molecule are replaced by other groups in a process known as nitration (Figure 3.1). To prepare cellulose nitrate, pre-dried cotton linters are treated with concentrated nitric and sulphuric acids. Water tends to slow the reaction and reduce the extent of nitration, so its presence in the reactants is minimized. Sulphuric acid catalyses the reaction, so it can take place under ambient conditions. The product is washed with water to remove residual acids. If residues of sulphuric acid are allowed to remain, an explosive reaction can occur. Remaining water in the cellulose nitrate is removed by washing it in alcohol.

The likelihood of a particular hydroxyl group being replaced is largely determined by its position in the molecule. Substitution of all three hydroxyl groups on each cellulose repeat unit results in the creation of explosive cellulose trinitrate commonly known as gun cotton, containing 14.4 per cent nitrogen. Cellulose dinitrates containing between 11 and 13 per cent nitrogen (an average degree of substitution between 1.9 and 2.7 hydroxyl groups) are useful for plastics, photographic film bases and lacquers. New cellulose nitrate is a colourless polymer which is brittle and tough, but flexible when plasticized and cast as thin films and sheets. Early plasticizers included oils and natural low molecular weight polymers, but the most commercially important early plasticizer was camphor, a white crystalline solid which can be extracted from the trunks, roots and branches of *cinnamomum camphora* trees in Asia and Florida or can be synthesized. Camphor sublimes (evaporates from the solid state) at room temperature and is known for its musty smell which deters moths and animals. Cellulose nitrate and camphor melts were used to produce transparent photographic and movie film bases known as celluloid, which revolutionized the film industry. Until the start of the twentieth century the film industry had relied on inflexible glass-film bases (Figure 3.2).

Cellulose acetate (CA)

Schutzenberger first attempted to acetylate cellulose in 1865, but found the reaction too difficult to control. Treatment with acetic acid and acetic anhydride resulted in replacement of all three hydroxyl groups in cellulose with acetyl groups ($-OCOCH_3$) to form the triacetate, a material with poor mechanical properties and only soluble in expensive chlorinated solvents. It was discovered

Figure 3.2

Lid of a box of cellulose nitrate hair combs (not brushes as it states on the box) showing the Xylonite factory and the dates of three patents taken out on celluloid. (source http://www.mernick.co.uk)

in the early 1900s that the more useful diacetate could be made in two stages. First, cellulose was acetylated with acetic anhydride and sulphuric acid to produce the triacetate, then partially hydrolyzed with dilute sulphuric acid to give the diacetate. Cellulose diacetate is soluble in inexpensive acetone. Cellulose acetate's most important commercial application was in non-flammable film bases. CA was used as a thermoplastic moulding material in the early part of the twentieth century and as a non-flammable dope applied to tighten and windproof linen aircraft wings during World War I. Today it is still used to make some hairbrushes and in the manufacture of filter tips for cigarettes.

Casein-formaldehyde or casein (CS)

The synthesis of casein-formaldehyde is different to the previously described semi-synthetic polymers because chemical treatment or formolizing of the natural polymer is applied after the shaping stage. Casein-formaldehyde is based on the casein protein protein in cow's milk. Casein is also present in vegetables including soya beans and wheat. Casein-formaldehyde plastic was first produced in 1897 by Spitteler and Kirsche by precipitating casein from milk using enzyme rennet (dried extract of rennin obtained from a cow's stomach). Today either acid coagulation or rennet coagulation techniques are used. The pH of fresh skimmed milk is adjusted to 6 and rennet added before heating from an initial temperature of 35–60°C to induce coagulation. The milk is stirred to produce protein in fine powder form. Casein particles are washed and excess water extracted by pressing prior to drying.

Since the 1950s, casein has been treated by the 'dry process' where the powder is plasticized with water before adding colouring agents and other additives. The resulting powder mixture is extruded using heat (30–65°C) to form rods, moulded shapes or sheets which are used to produce knitting needles, buttons and pen barrels. The forms are formolized by being immersed in a 4–5 per cent aqueous formaldehyde solution at pH 4–7 and held at 16°C for a period of 2 days to 3 months depending on the thickness of the piece. Formolizing introduces crosslinks between the peptide groups of the casein protein (—CONH—) to harden and improve its resistance to swelling in moisture. The casein-formaldehyde is dried slowly.

3.1.2 Preparation of synthetic polymers

The process of chemically joining low molecular weight monomers to form a high molecular weight macromolecule is known as polymerization and occurs via one of two major reaction types, known as *addition polymerization* and *condensation polymerization* (Brydson, 1999). Most polymerization reactions do not progress to completion, resulting in the residues of starting materials in the resulting plastic. Approximately 1–3 per cent of residual monomer is found in acrylics, poly (vinyl chloride), polystyrene, polycarbonates, polyesters, polyurethanes and formaldehyde polymers immediately after production. Monomers with boiling points lower than ambient are likely to have evaporated before the final product is used, while those with higher boiling points, including styrene monomer and terephthalates used in polyesters, offgas slowly from the plastic formulation and can often be detected by odour.

In the next section, the principles of addition reactions are first outlined followed by details of the synthesis processes for the addition polymers most commonly found in heritage collections. Condensation polymerization is presented using the same format.

Addition polymerization, also included in the modern term chain-growth polymerization, is a process in which unsaturated monomer units (those containing at least one double or triple carbon-carbon bond) are joined together one at a time to produce a linear polymer. Addition polymerization can be visualized as threading beads (monomers) on a string one by one. The resulting polymer contains only the small molecule which polymerizes the repeat unit found in the monomers and no by-products are formed. The simplest useful addition polymer is polyethylene which can be described as:

$$-CH_2-CH_2-CH_2-CH_2-CH_2-CH_2-CH_2-CH_2-CH_2-\text{or } (-CH_2-)_n$$

where n is approximately 20 000 and a typical length of chain is $3\,\mu$ with a diameter of around 1 nm. The lengths of polymer chains are not all identical, which results in a distribution of molecular weights for each material. Although the covalent bonds linking the repeat units together, known as backbone links, are illustrated as if they are aligned, collinear and single, this is a simplification and the angle and type of bonds varies between polymers.

Polymers formed by addition polymerization include polyethylene, polypropylene, poly (methyl methacrylate), polystyrene and poly (vinyl chloride). These are known as *homopolymers* because only one type of monomer is used as starting material. The linear structure of polymers created by addition polymerization reactions imparts properties such as the ability to be repeatedly softened on heating (thermoplasticity), high flexibility and good solubility in hydrocarbons.

Copolymers are polymers produced from more than one monomer with the aim of forming a polymer which exhibits a combination of the properties of all starting materials. Copolymerization of vinyl acetate with vinyl chloride improves the flexibility and processability of PVC which otherwise tends to degrade on heating. Vinyl singles and long playing (LP) records are made from a vinyl acetate/vinyl chloride copolymer. Copolymerization of styrene (softening temperature of polystyrene is 100°C) and methyl methacrylate (softening temperature of poly [methyl methacrylate] is 120°C) improves polystyrene's temperature resistance. The properties of a copolymer can be varied by adjusting the ratio of component monomers, with the monomer in highest proportion having more influence on the final product than the monomer present to a lesser extent.

Addition polymerization

Addition polymerization can be achieved via several different types of mechanism. The *free radical mechanism* was one of the first identified. Free radicals are very reactive atoms or molecules which have a spare unpaired electron. Electrons need to be in pairs to achieve stability. The free radical mechanism occurs in three distinct stages known as initiation, propagation and termination. During initiation, free radicals are generated by the reaction between a catalyst and monomer. Commonly used catalysts include benzoyl peroxide and azodi-isobutyronitrile, which are encouraged to produce reactive free radicals by exposing them to light or heat. The free radicals react with monomer molecules, generating even more radicals which react further with monomer molecules (Figure 3.3). Unsaturated bonds are broken during initiation and the released bonds are free to attach to other monomer molecules.

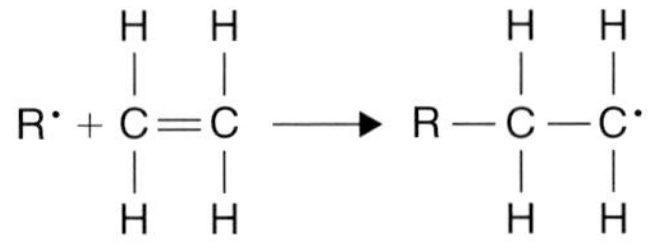

Figure 3.3
Initiation stage of free radical mechanism for polyethylene showing generation of reactive free radicals. 'R•' represents the active initiator and '•' is an unpaired electron.

In the second stage known as propagation, monomers continue to add together linearly, forming increasingly longer chains (Figure 3.4). Chain growth is rapid and 1000 units join in around 0.001 seconds (Callister, 1994).

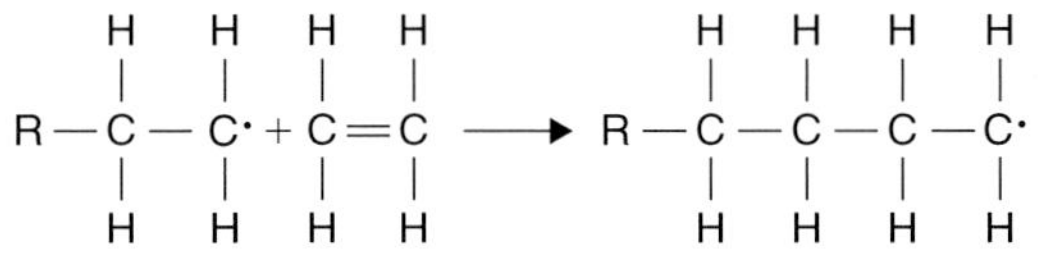

$$R-\overset{\overset{\displaystyle H}{|}}{\underset{\underset{\displaystyle H}{|}}{C}}-\overset{\overset{\displaystyle H}{|}}{\underset{\underset{\displaystyle H}{|}}{C^{\textbf{·}}}} + \overset{\overset{\displaystyle H}{|}}{\underset{\underset{\displaystyle H}{|}}{C}}=\overset{\overset{\displaystyle H}{|}}{\underset{\underset{\displaystyle H}{|}}{C}} \longrightarrow R-\overset{\overset{\displaystyle H}{|}}{\underset{\underset{\displaystyle H}{|}}{C}}-\overset{\overset{\displaystyle H}{|}}{\underset{\underset{\displaystyle H}{|}}{C}}-\overset{\overset{\displaystyle H}{|}}{\underset{\underset{\displaystyle H}{|}}{C}}-\overset{\overset{\displaystyle H}{|}}{\underset{\underset{\displaystyle H}{|}}{C^{\textbf{·}}}}$$

Figure 3.4

Propagation stage of free radical mechanism for polyethylene. Monomers join together to form long chains.

Addition polymerization processes can terminate in various ways. The most common method of termination is for the active ends of two chains to react with each other thus preventing further growth.

Free radical polymerization takes place at high temperatures and pressures (around 300°C and 2000 atmospheres) and the activity of the free radicals is difficult to control. Uncontrolled reaction and longer chain lengths during propagation increase the chances for branches to form in the linear polymer chain. The chain curls back on itself and breaks, leaving irregular chains sprouting from the main polymer backbone. Branching reduces the density of the polymer and results in lower tensile strength and lower melting points compared with purely linear polymers. In addition, the random nature of the termination step results in polymer chains of varying lengths.

A technique developed in the 1950s using Ziegler-Natta catalysts (for example, triethylaluminium in the presence of a metal chloride) allows more control than the traditional free radical method. The monomer sits between the aluminium atom and an ethyl group in the catalyst. The polymer then grows out from the aluminium atom and has fewer chances to branch. Other alternatives to free radical polymerization for addition polymers include cationic and ionic polymerization. These techniques are not used extensively in industry, because their stringent reaction conditions include absence of water and oxygen which is complex and expensive to achieve, but are employed to produce those polymers which cannot be made by free radical techniques, including polypropylene.

Addition polymerization may be carried out in bulk so that monomers react while in solution or the monomers may be first dispersed in suspension or emulsion. Polymerizing in bulk increases the risk of reactants to overheat and induce an uncontrolled reaction. Heat is generated because polymerization reactions tend to be exothermic and polymers are poor thermal conductors. Stirring of reactants acts to disperse the generated heat. Another challenge is to recover excess solvent following polymerization. This is necessary to reduce pollution of the surrounding area, to limit the risk of fire and for cost reasons.

Polymerization in suspension involves stirring the monomers vigorously into water where they disperse into droplets. A suspension agent such as gelatine or talc is added and coats each droplet to prevent it forming large drops again.

Polymerization reactions take place within each droplet. In emulsion polymerization, monomers, water, initiator and soap (5 per cent by weight of mixture) are stirred together (Brydson, 1999). The monomer forms droplets which are surrounded and protected by soap molecules while polymerization takes place. Emulsion polymerization is a low-polluting, low-flammability technique but soap residues adversely affect the appearance and electrical properties of the final product.

Chemical structures for the repeat units and IUPAC names for monomers of the most industrially important synthetic addition polymers are shown in Table 3.1.

Table 3.1   Chemical structures for the repeat units of synthetic addition polymers

Industrial name of polymer	Polymer acronym	Monomer (industrial name)	Monomer (IUPAC name)	Repeat unit structure
polyethylene	PE	ethylene	ethene	$+CH_2-CH_2+$
polypropylene	PP	propylene	propene	$+CH_2-CH+$ $\quad\quad\quad CH_3$
polybutadiene	BR	butadiene	but-1-ene-1,4-diyl 1,4-butadiene	$H \quad\quad CH_2+$ $\quad C=C$ $+CH_2 \quad\quad H$
polystyrene	PS	styrene	phenylethene	$+CH_2-CH+$ (phenyl ring)
poly (methyl methacrylate)	PMMA	methyl methacrylate	methyl (2 methyl propenoate)	CH_3 $+CH_2-C+$ $\quad\quad C=O$ $\quad\quad O$ $\quad\quad CH_3$
polytetrafluoroethylene	PTFE	fluoroethylene	1,1',2,2'tetrafluoroethene	$F \quad F$ $+C-C+$ $F \quad F$
poly (vinyl chloride)	PVC	vinyl chloride	chloroethene	$+CH_2-CH+$ $\quad\quad\quad Cl$

Polyethylene (PE)

Polyethylene (also known as polythene) was synthesized by accident in 1932 when scientists at Imperial Chemical Industries (ICI) investigated the reaction between ethylene and various compounds at high pressure. Polyethylene is generally commercially polymerized from monomer ethylene gas under high pressure (1000–3000 atmospheres) and at temperatures of 80–300°C (Brydson, 1999). A free radical initiator such as benzoyl peroxide, azodi-isobutyronitrile or oxygen is added. The reaction is exothermic and must be cooled to control the rate of polymerization and molecular weight.

Polyethylene exists in several forms which vary in the extent and length of branching and molecular weight and are dependent on the pressure and temperature conditions applied during synthesis. Commercial low density polyethylene (LDPE) exhibits between 40–150 short side chains for every 1000 ethylene units. LDPE is produced by polymerizing ethylene at high pressure (1020–3400 atmospheres) and 350°C. It has a density of 0.912–0.935 gcm^{-3}. High density polyethylene (HDPE) is produced using catalysts such as the Ziegler-Natta catalysts described earlier in this chapter and has between 1 and 6 short side chains per 1000 ethylene units. HDPE has a density approximately 0.96 gcm^{-3} and exhibits greater chemical resistance, hardness, stiffness and tensile strength than LDPE.

Polypropylene (PP)

Polypropylene was first polymerized in 1954 from propylene gas using Ziegler-Natta catalysts. Polypropylene is made commercially using the suspension process at around 60°C and conversions of 80–85 per cent from monomer to polymer are achieved (Brydson, 1999). The reaction mixture is centrifuged to recover solvent and unreacted catalyst. The polymer is washed and dried at 80°C prior to blending it with antioxidants and extruding it into pellets.

Polystyrene (PS)

The discovery of polystyrene is generally attributed to E. Simon in 1839 who believed he had made an oxidation product of styrene, but it is likely to have been discovered earlier (Kaufman, 1963). Bulk polymerization technique is the most widely used today to produce commercial PS. Styrene is partially polymerized at 80°C before being run into a tower fitted both with heating and cooling facilities. The top of the tower is maintained at around 100°C while the bottom is heated to 180°C. The higher temperatures at the base of the tower boil off any unreacted styrene from the polymer. Freshly made, liquid PS is directed into an extruder where it is shaped and allowed to cool.

Acrylonitrile-butadiene-styrene (ABS) copolymer

ABS copolymers were first commercially available in 1948 (Mossman, 1997). The most widely used method to prepare ABS is to add styrene and acrylonitrile monomers to polybutadiene latex and warm the mixture to around 50°C to allow absorption of the monomers. An initiator is then added to polymerize the styrene and acrylonitrile. A mixture of polymeric products is obtained including styrene-acrylonitrile.

Poly (methyl methacrylate) (PMMA)

PMMA was first produced commercially in the UK and in Germany in the early 1930s. Monomer methyl methacrylate polymerizes readily under ambient conditions and is therefore supplied with an inhibitor (up to 0.1 per cent of hydroquinone). After removal of the inhibitor, free radical polymerization with peroxides or azodi-isobutyronitrile at 100°C is employed commercially. Oxygen slows the rate of polymerization and leads to the formation of peroxides so is excluded by allowing nitrogen into the reaction vessel. Shrinkage between monomer to polymer is high at around 20 per cent.

Polytetrafluoroethylene (PTFE)

PTFE was first produced in large scale in 1950 (Kaufman, 1963). There is little published information containing details of the commercial polymerization process for PTFE, but the product is known to form either as granules or dispersed low molecular weight particles. Polymerization is by free radical reaction at temperatures of 80–90°C and gives a yield around 85 per cent (Brydson, 1999).

Poly (vinyl chloride) (PVC)

Both unplasticized or rigid and plasticized PVC were commercially available around 1940, although applications for the plasticized material developed more rapidly than for the rigid material. Vinyl chloride is polymerized via free radical reaction. The molecular weight of the resulting PVC is determined by the polymerization temperature, which ranges from 50–75°C. The reaction is highly exothermic. The heat of polymerization is 92 kJ/mol, which is higher than that for polystyrene (70 kJ/mol) and poly (methyl methacrylate) (56.5 kJ/mol) (Roderiguez, 1989). Eighty per cent of commercial PVC polymers are produced by suspension polymerization. The resulting PVC particles are roughly spherical and range from 50–250 μm in diameter. The degree of polymerization or number of repeat units in the molecular chain, n, ranges between 500 and 1500, which corresponds to a theoretical molecular weight average of 100 000–200 000. In practice, all batches contain molecules with a range of chain lengths.

After synthesis, addition polymers are either in solution or are dried to powder, pellets or flakes. To convert polymers to plastics, it is necessary to incorp-orate additives by stirring or grinding them in at high speed or by heating polymer and additives together prior to shaping.

Condensation polymerization

Condensation polymerization is also included in the modern term step-growth polymerization. It comprises stepwise intermolecular reactions that usually involve more than one type of molecular species incorporating alcohol, amine or carboxyl functional groups. When the species join together, they exclude or condense a small molecule which may be water, alcohol or acid. The monomers do not have the same chemical formula as the repeat units (Figure 3.5). To control the rate of polymerization and the chemical structure of the final polymer, it is usual to build the polymer in steps. A prepolymer or oligomer of medium molecular weight is formed first and is further polymerized to a high molecular weight final product.

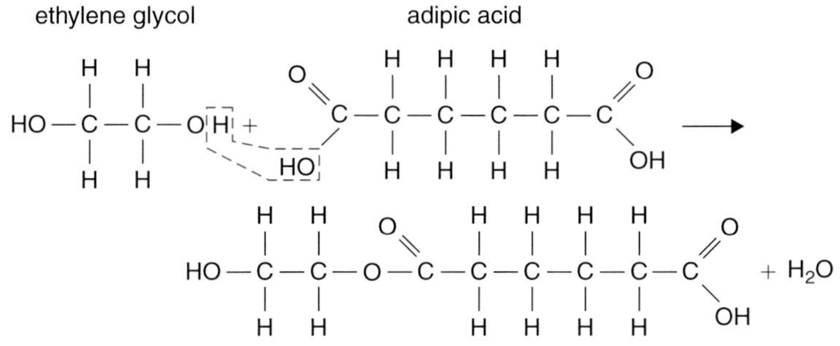

Figure 3.5

Condensation polymerization usually involves more than one monomer species and a small molecule such as water is excluded on reaction. Polyesters are formed by condensation reaction between a glycol and an acid as shown here. The reaction takes place many times to form a high molecular weight polymer.

Reaction rates are slower than for addition polymerization and lower molecular weight products are formed. Condensation polymers include polyesters, nylons, polycarbonates, formaldehyde polymers and epoxies. Compared with polymers created by addition polymerization, products of condensation reactions harden on heating (thermosetting), are brittle, poorly soluble in hydrocarbons and may be swollen but not dissolved by chlorinated liquids. More details of the synthesis processes for the condensation polymers most commonly found in heritage collections are presented next.

Phenol-formaldehyde (PF)

Sir Gilbert Morgan discovered phenol-formaldehyde polymers in the early 1890s while attempting to make artificial dyestuffs, but it was Leo Baekeland in the USA that exploited the reaction and obtained a patent for a commercial synthetic polymer in 1907. Phenol and formaldehyde were reacted in alkaline conditions with loss of water which formed an intermediate solid or liquid product. At this stage the product comprises linear molecules and is soluble. When heated under pressure and with acid to 150°C, the intermediate is converted to crosslinked, hard, solid polymer Bakelite (Figure 3.6).

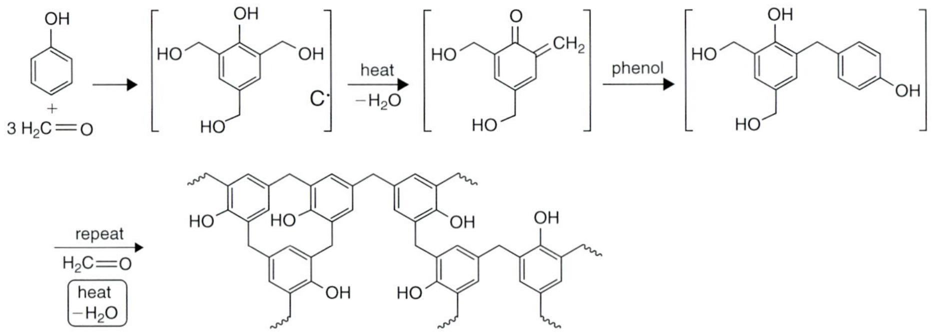

Figure 3.6

Synthesis of Bakelite (phenol-formaldehyde) by condensation polymerization.

More recently, the prepolymer has been made by reacting phenol and formaldehyde in acidic conditions to make a novolak. Novolaks are also crosslinked and hardened by heating with acid (Brydson, 1999). Because crosslinking is an exothermic process which causes shrinkage of the polymer, inert fillers such as cellulose fibres or fine sawdust from softwoods are incorporated prior to moulding. In addition to reducing shrinkage, fillers also improve the physical properties of PF plastics by increasing their resistance to impact. Unfilled PF polymers are hard, brittle materials and fillers increase their flexibilities (Horie, 2002). In common with other condensation polymers, PF polymers are thermosetting so can only be heated and moulded once.

Urea-formaldehyde (UF)

Commercial interest in urea-formaldehyde polymers as adhesives and waterproofing agents for textiles originated in 1918 with a patent by John while UF moulding powders were first developed in 1926 and are still prized today for their excellent insulation properties (Brydson, 1999). Urea, synthesized from ammonia and carbon dioxide at 200°C and 300 atmospheres pressure is heated to 40°C aqueous solution of formaldehyde in the ratio 1:1.3 to 1:1.5 under neutral or mildly alkaline conditions (Figure 3.7). Unlike phenol-formaldehyde polymers, UF polymers are not themselves deeply coloured but have a light appearance. Bleached wood pulp is used to fill the prepolymer formed at this stage with a ratio 2:1 prepolymer:filler (Brydson, 1999). The filled prepolymer is dried and ground to powder. Large quantities of pigments are added to counteract the brown colouration imparted by the wood pulp. The resulting prepolymer is crosslinked into a 3D network with heat (125–160°C) during the moulding stage. Crosslinking imparts high resistance to organic solvents, though attack by acids and alkalis is possible.

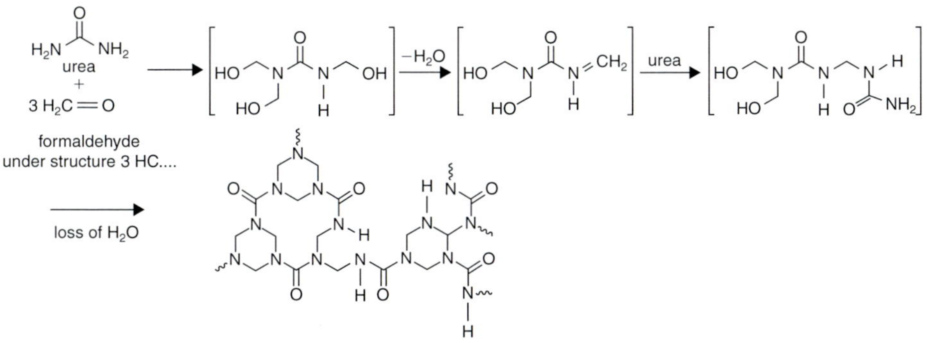

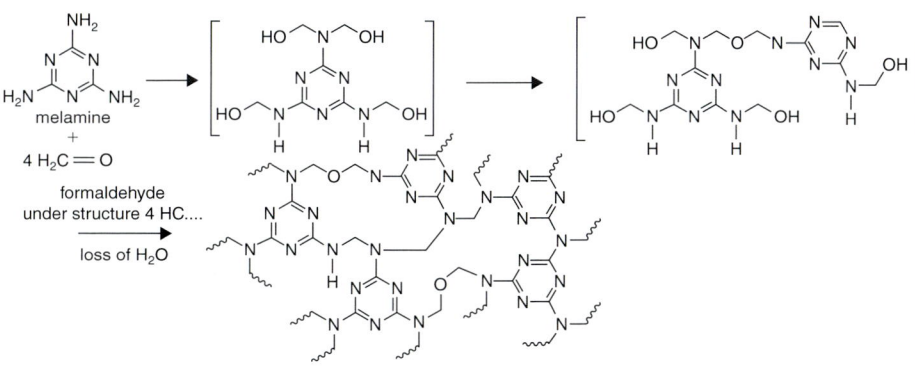

Figure 3.7

Synthesis of urea-
formaldehyde
by condensation
polymerization.

Melamine-formaldehyde (MF)

The production of MF was first patented in 1935 by Henkel (Brydson, 1999). Today, MF polymers have commercial applications in decorative laminates for chipboard (a low cost substitute for solid wood) and unbreakable tableware. During synthesis, an excess of formaldehyde is heated with melamine at 80°C in alkaline conditions to produce an aqueous, syrup-like prepolymer. The prepolymer contains melamine-formaldehyde in a molar ratio of 1:2 (Figure 3.8). It is compounded with fillers, pigments, lubricants, stabilizers and sometimes accelerators, dried and milled to powder. Heating the powder to mould it at 145–165°C and 30–60 MPa produces a solid, thermosetting network.

Figure 3.8

Synthesis of melamine-
formaldehyde
by condensation
polymerization.

Nylon polymers (PA)

Nylons belong to a class of polymers known as polyamides. Nylons are distinguished from each other by a numbering system based on the number of carbons in the starting materials. Nylon 66, which was first prepared in 1935, was used to make stockings and is still the major commercial nylon. It is prepared by reacting hexamethylene diamine (6 carbon atoms) with adipic acid (6 carbon atoms) via a condensation reaction (Figure 3.9). Nylons can also be synthesized from a single starting material, such as a long-chain amino acid like aminodecanoic

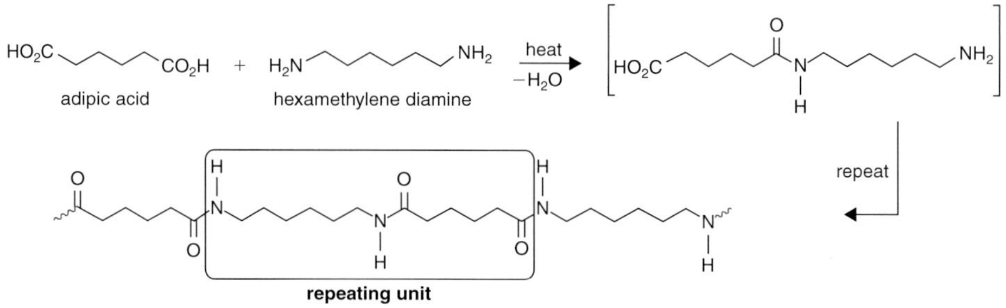

Figure 3.9

Synthesis of Nylon 6,6 by condensation polymerization.

acid (11 carbon atoms) which undergoes self-condensation to produce Nylon 11. The synthesis of nylon does not require an external catalyst, but is catalysed by the acid starting materials. The reaction mixture is heated to 200–265°C for up to 12 hours and oxygen is excluded by flushing the reaction vessel with nitrogen. Water is excluded as the reaction proceeds.

All nylons show high impact strength, toughness, flexibility and abrasion resistance. They are highly resistant to solvents but are attacked by concentrated acids at room temperature. Nylons are used in clothing, brushes, surgical sutures, bearings and gears.

Polycarbonate polymers (PC)

Polycarbonates were first prepared by Einhorn in 1898 but were not produced commercially in Germany and the USA until 1958 (Brydson, 1999). They are used as alternatives to glass in greenhouses, in roofing as fibres and as the support material for CDs and DVDs. Reaction of polyhydroxy compounds with polybasic acids produces polymers with ester groups (—COO—) by eliminating water, known as polyesters. When carbonic acid is used as one of the starting materials, the resulting polymer contains carbonate (—O.CO.O—) linkages and is known as a polycarbonate. Carbonic acid is not a stable, free compound, so must be generated via another reaction. Polycarbonates may be synthesized by reacting phosgene with bisphenol A (2,2-bis [4-hydroxyphenyl] propane). The reaction is a condensation type with hydrogen chloride being excluded (Figure 3.10).

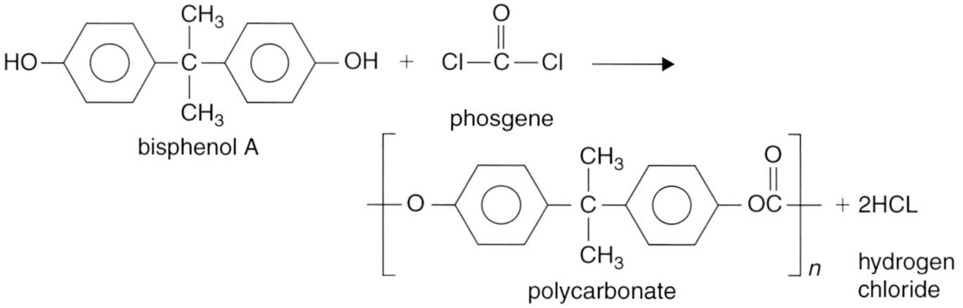

Figure 3.10

Polycarbonate polymers are formed via a condensation reaction to make a high molecular weight product comprising n repeat units.

Polyurethane polymers (PUR)

Basic polyurethane chemistry was discovered by Otto Bayer in 1937, but polyurethane polymers were first developed as replacements for rubber at the start of World War II. Numerous applications followed including fibres, rigid and flexible foams, mouldings and elastomers (Brydson, 1999). The preparation of polyurethane polymers occurs via a reaction process intermediate between those of addition and condensation (Brydson, 1999). Like addition polymerization, there is no splitting off of small molecules, but the kinetics are otherwise similar to condensation polymerization.

Polyurethane polymers are products of a polyol, based either on a polyester (product is PURester) or polyether (product is PURether), with several alcohol groups (—O—H), a di-or poly-isocyanate with several cyanate groups (—N═C═O), and a chain extender. The chain extender reacts with the polyol's alcohol groups, initiating an imbalance in negative and positive charges throughout the molecule which, after reaction with the isocyanate, results in the formation of a urethane group (—NHCOO—) as shown in Figure 3.11. Isocyanates are highly toxic. PUResters were the earliest foams and have superior mechanical properties but poorer chemical resistance than the more recent PURethers. The physical properties of polyurethanes are determined by the chemical formulae of both the isocyanate and the alcohol which control the molecular weights and extent of crosslinking in the resultant polymer.

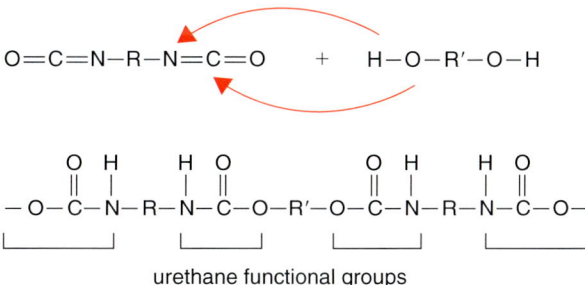

urethane functional groups

Figure 3.11
The urethane link in polyurethane polymers is formed by reaction between cyanate and alcohol groups.

To produce polyurethane foam, water is added to the polyol and polyisocyanate starting materials. The water molecules react with the cyanate groups to form amine groups (—NH$_2$) and carbon dioxide gas. The amines continue to react with isocyanate groups to form urea linkages (—HNCONH—) between the chains, instead of the urethane groups created in the absence of water. As polymerization progresses accompanied by an increase in molecular weight, carbon dioxide gas becomes trapped in the increasingly viscous liquid polymer. The trapped bubbles create cells in the polyurethane foam as it hardens.

Epoxy polymers (EP)

Epoxy polymers (also known as epoxides or epoxylines) were first developed in the 1930s by Pierre Castan and were commercially produced in 1939 (Mossman, 1997). The high early production costs compared with those of polyesters have limited their applications but they are established as protective coatings, laminates and construction materials.

Epoxy polymers comprise two components. A liquid, low molecular weight, diepoxy prepolymer is prepared and is then crosslinked with a low viscosity diamine hardener. The diepoxy is formed by reacting bisphenol A with epichlorohydrin in the presence of sodium hydroxide to form a molecule resembling two molecules of bisphenol A but with epoxy groups replacing the hydroxyl (OH) groups at each end (Figure 3.12). The nitrogen atoms in the amine groups (NH_2) present in the hardener react with the carbon atoms in the epoxy group. This reaction links the diepoxy molecules together to form a crosslinked or cured network.

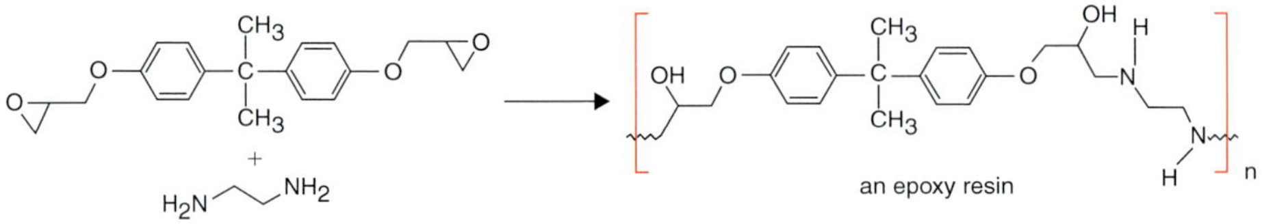

Figure 3.12

The diepoxy prepolymer (upper left structure) reacts with an amine hardener (lower left structure) to form an epoxy polymer.

A wide range of physical properties can be achieved by varying the molecular weight of the epoxy component, the type of hardener and by adding catalysts and plasticizers. Epoxy polymers can be reinforced with glass, carbon and synthetic fibres prior to curing to increase strength and flexibility.

Epoxies shrink by approximately 5 per cent between mixing prepolymer with hardener and the formation of a networked, hard polymer (Horie, 2002). In order to achieve effective crosslinking, the number of epoxy groups must be matched with a suitable number of amine groups in a particular quantity, known as the chemical equivalent. The type of hardener determines the speed of curing, the temperature of curing and the amount of heat generated by the curing reaction. Diethylenetriamine and triethylenetetramine are highly reactive amines which cure epoxy prepolymers at room temperature within one hour. The disadvantage with such a short pre-cure period or pot life is the brief period available to manipulate the material, both for industrial and artistic projects. The curing reaction is exothermic and the mixture may attain 250°C. Such temperatures thermally age the epoxy which results in the formation of a yellow colouration on curing (Figure 3.13). Other hardeners including dimethyl-aminopropylamine exhibit a pot life of more than two hours.

Figure 3.13

Handbag sculpture by Danish artist, Mette Ussing, made in 2005. The work was cast in Epoxy BK® which has a short pot life and attains temperatures up to 175°C on curing. The result is trapped air bubbles, visible on the rear side of the sculpture, and yellowing.

Polyester polymers

Polyester polymers are materials which exhibit various chemistries but all contain ester linkages in the polymer chain. They have attained industrial importance as moulded materials, fibres, packaging film and as structural materials which are reinforced with fibres and fillers. Chemists, John Rex Whinfield and James Tennant Dickson, employees of the Calico Printer's Association of Manchester, patented polyethylene terephthalate in 1941, after advancing the early research of Wallace Carothers. Poly (ethylene terephthalate) is the basis of polyester fibres and fizzy drinks bottles. The first polyester fibre known as Terylene was also developed in 1941.

The main types of polyesters found in museums may be divided into saturated and unsaturated. Saturated polyesters are hard, crystalline, strong, and tough. They are commonly used in fizzy drinks bottles, as well as in magnetic tape for video, audio, and computers. They are also used in X-ray film, strapping, labels, and packaging. They are thermoplastic, so repeatedly soften when heated and harden when cooled. Unsaturated polyesters are used to make glass-reinforced plastics and were first widely used in the USA during World War II. They are used to make boats, exterior building structures, luggage, tennis rackets and fishing rods. Unsaturated polyesters are thermosets, so can only be softened once by warming before they deteriorate.

Saturated polyesters such as poly (ethylene terephthalate) (PET) are condensation polymers. PET is the product of an esterification reaction between ethylene glycol and terephthalic acid with the exclusion of water molecules (—H from alcohol and —OH from acid) (Figure 3.14).

Unsaturated polyesters are formed via a condensation reaction between dicarboxylic acids and dialcohols (glycols). A viscous liquid is produced by

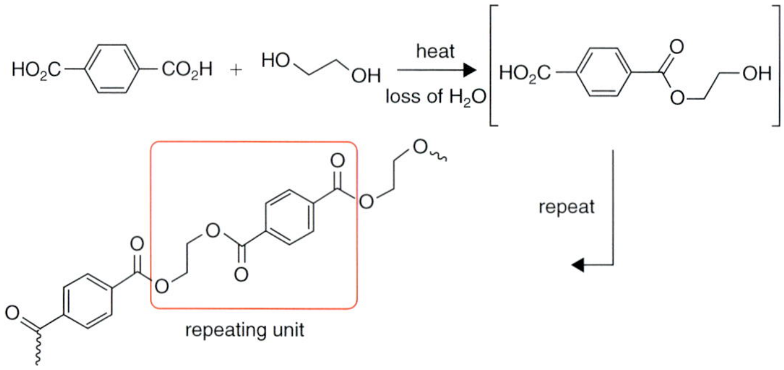

Figure 3.14

Polyesters are formed from esterification between a polyol and acid with the exclusion of water.

repeating unit

heating the components at 200–210°C for up to 12 hours with constant stirring and under nitrogen to control the rate of polymerization. The presence of nitrogen also prevents discolouration due to oxidation. Various dicarboxylic acids and glycols produce polyesters with a range of properties. The most commonly used dicarboxylic acids are maleic anhydride, maleic acid and fumaric acid. Dicarboxylic acids introduce carbon-carbon double bonds into the polyester. Water is formed by the condensation reaction and removed.

The polymer is cooled, dissolved in styrene monomer and a catalyst added, typically cobalt naphthenate. Crosslinking reactions between the carbon-carbon double bonds in the polyester polymer and styrene lead to the formation of a three-dimensional network and are initiated by adding a peroxide. Peroxide and catalyst must not mix as they form an explosive combination. The heat generated by crosslinking can result in distortion and discolouration. Heating can be minimized by reducing quantities of accelerator and catalyst. Crosslinking results in shrinkage of up to 8 per cent by volume, which can be reduced if heating is minimized. After crosslinking, between 2–4 per cent styrene monomer remains. High concentrations of filler are added to polyesters used to make reinforced materials to compensate for the high shrinkage on curing. Unsaturated polyesters can be reinforced with glass, carbon and synthetic fibres prior to curing to increase tensile strength. The inclusion of glass fibres in polyesters are thought to protect the polymer against deterioration by water, a useful property when glass-reinforced polyesters are used to build and repair boats and cars (Aurer and Kasper, 2003).

3.2   Additives

Both chemical and physical performances and the longevity of polymers can be changed dramatically by incorporating additives during processing. Additives for plastics formulations, also known as modifiers, comprise chemically diverse materials ranging from complex organic molecules used as antioxidants and light stabilizers to simple inorganic compounds including talc (hydrated magnesium silicate), which imparts opacity and increases tensile strength.

The type and quantity of additives permit many different products to be made from one polymer. For example, raw PVC polymer is a brittle, inflexible material with rather limited commercial applications. Attempts to process raw PVC using heat and pressure result in severe degradation of the polymer. Compounding PVC involves incorporating sufficient additives into the raw polymer to produce a homogeneous mixture suitable for processing and with the required final properties for the lowest possible price. PVC can be used to produce flexible toys, electric cable insulation, photograph pockets and shoe soles simply by varying the amount of plasticizer in the formulations between 16–50 per cent by weight.

With the exception of particulate fillers, additives are the most expensive components of plastics formulations. For this reason, sufficient amounts of each additive are included in formulations so that the final products have exactly those properties and useful lifetime for which they were designed, but not in excess. The useful lifetime is dependent on the application, polymer type and cost (Table 3.2). Because plastics in cultural collections are often required to have longer active lifetimes than those for which they were designed, many of their additives are either exhausted during pre-collection use or post-collection storage or display.

Table 3.2 Average lifetime for plastics in Europe (European Union Commission, 2000)		
Category	**Example of application**	**Average lifetime (years)**
building	window frames	10–50
electric and electronic	cable insulation	21
furniture	upholstery covering and foam	17
automotive	dashboard	12
household appliances	housings for food mixers, vacuum cleaners, etc.	11
medical	blood bags	2–10
packaging	bags and film	1

Polymer additives or modifiers may be generally grouped into those which alter physical properties, those which change chemical properties and those which affect the appearance of the finished plastic product.

Table 3.3 summarizes the types and function of the additives which are most commonly incorporated into polymers. There is some overlap between the properties and actions of many polymer additives. For example, plasticizers

Table 3.3 Summary of types, functions, examples and concentrations of the most frequently used additives

Additives	Major functions	How achieved	Examples	Approximate concentration in plastics formulation by weight (%)
plasticizer	to soften polymer; reduce glass transition temperature	separates polymer chains from each other	phthalate esters; aliphatic diesters; epoxidized oils; phosphate esters; polyesters	20–50
filler	to opacify; increase hardness; reduce cost; reinforce polymer	changes refractive index and reflective properties; adds bulk; diverts stress from polymer	calcium carbonate; magnesium carbonate; barium sulphate; glass, carbon and nylon fibres	0–20
lubricant	to prevent adhesion of tacky polymer to metal processing equipment	sweats out due to low compatibility with polymer	calcium stearate; normal and dibasic lead stearate	<1
impact modifier	to increase impact strength	toughens compound	styrene-acrylonitrile; ethylene-vinyl acetate copolymer	2–5
antioxidant	to retard oxidative degradation	interrupts free radical reaction sequence by binding oxygen or degradation product so it is unavailable for further reaction	arylamines;phenolics (butylated hydroxytoluene, BHT); organophosphites; metal deactivators; carbon black	0–0.2
heat stabilizer	to protect against thermal degradation	inhibits degradation reactions; stabilizes thermal degradation products	barium-zinc, barium-cadmium and lead octoates; naphthenates and benzoates; epoxidized soya bean and linseed oils	>1
light stabilizer	to protect against degradation due to UV radiation	absorbs UV radiation; quench degradation reactions	carbon black; benzo-phenones; benzotriazoles; salicylates; hindered amine light stabilizers (HALS)	>1
flame retardants	to inhibit ignition, smoke generation and rate of burning	minimizes access to polymer, heat or oxygen	aluminium trihydrate; magnesium hydroxide; phosphates; chlorinated paraffins	2–60
blowing agents	to introduce gas into liquid polymer to produce foam	gas, evaporated solvent or chemical reaction on solid to generate CO, CO_2 or N_2	CO_2, N_2 and air; aliphatic and halogenated hydrocarbons; azocarbonamide	varies
colouring agents	as decoration; protection from radiation	adds colour and opacity to surfaces or to bulk	carbon black; inorganic pigments; organic dyes; e.g. anthraquinones	varies

may also act as impact modifiers, lubricants and stabilizers. Conversely, a commercial lubricant may act as a plasticizer. The major requirements for additives in plastics include:

- stability under processing temperatures and pressures
- stability when the plastic is in use
- ability to remain in the formulation throughout the intended useful lifetime of the plastic and not to migrate out or evaporate. The intended useful lifetime of most plastics is between 1 year (polyethylene carrier bag) and 50 years (unplasticized PVC window frames)
- low toxicity and inert odour or taste, especially in plastics for use with food or medicines
- low cost

3.2.1   Additives which affect the physical properties of polymers

Additives which alter physical properties either during processing or in the final plastic include plasticizers, fillers, lubricants and flow promoters, impact modifiers and foaming agents.

Plasticizers

Plasticizers comprise the largest volume additives in today's plastics industry. Their primary functions are to impart flexibility, softness and extensibility to rigid polymers. In addition, plasticizers facilitate processability by reducing viscosity and lubricating thus allowing a lower processing temperature. Plasticizers also increase impact resistance and reduce the brittleness of plastics, particularly at low temperatures (Table 3.4). General purpose plasticizers

Table 3.4   Effect of plasticiser DEHP (di [2-ethylhexyl] phthalate) concentration on physical properties of PVC moulded at 178°C and under pressure for 5 minutes (Wilson, 1995)

DEHP (per hundred parts PVC, phr)	DEHP (% by weight based on PVC plus DEHP)	100% modulus[1] (N/mm^2)	Elongation at break[1] (%)	Low temperature flex point[2] (°C)
30	23.1	20.2	270	+3
40	28.6	14.7	300	−10
50	33.3	10.2	325	−22
60	37.5	7.0	370	−30
80	44.4	3.7	420	−44
100	50.0	2.4	430	−51

[1]As determined using BS 2782: Method 301E: 1970
[2]Lowest temperature at which a standard strip can be deflected through an angle of 200° under fixed torque without failing (BS 2782: Part 1: Method 150B:1976)

are high boiling, non-volatile solvents with molecular weights of at least 300, that is 100 to 1000 times lower than that of polymers, although polymers can also be used as plasticizers for specialist applications (Brydson, 1999). Because of the difference in polarity and weak interaction forces between polymer types, matching of plasticizer and polymer is necessary.

Plasticizers include esters of aromatic and aliphatic acids and anhydrides, epoxidized oil, phosphate esters, hydrocarbon oils and polymeric materials. PVC is the polymer most in need of plasticizers, but polyvinyl acetate, epoxies, cellulose nitrate and acetates also require these additives.

The stages of the plasticization process involve mechanically milling the solid, agglomerated (large particle size) polymer with sufficient liquid plasticizer to reduce the particle size and inhibit reagglomeration of the polymer particles. During this process, plasticizer molecules are effectively dispersed throughout the polymer and attach themselves to the surfaces of the polymer particles physically. Weak interaction takes place between the plasticizer and polymer particles. Such interactions have energy values in the order of 20 kJ/mol, approximately 10 times lower than such values for covalent, chemical bonds.

Next, the remaining plasticizer is added and the mixture heated. Plasticizer diffuses within particles. The plasticizer is probably present as clusters of molecules between bundles of polymer segments or molecules (Figure 3.15). The polymer chains are no longer held rigidly together, but are dispersed by the plasticizer molecules. On a molecular level, plasticization is the weakening or selective breaking of bonds between molecules, while leaving others intact, to increase intermolecular space, known as free volume. It is this increased space which allows room for changes in shape, flexing or moulding of the final material.

Figure 3.15

Plasticizer molecules (◄) diffuse between polymer chains and are found only in the areas where chains are loosely packed.

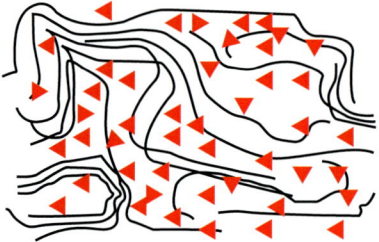

The first plasticizers for polymers were added to expand their commercial applications by reducing shrinkage during shaping and increasing flexibility. The applications for cellulose nitrate were limited, until Parkes added castor oil to it in 1865, and were further expanded when Hyatt added camphor in 1870 (Carraher, 2000). Camphor is a white, waxy solid with a pungent odour and a melting point of 180°C and a boiling point of 204°C. Tricresyl phosphate acts as both a plasticizer and flame retardant, so was used to replace part of the camphor in cellulose nitrate in 1910.

Of the one million tonnes of plasticizers used annually in Europe today, approximately 90 per cent comprise phthalate esters. Phthalate esters are manufactured by reacting phthalic anhydride with an alcohol. The most frequently used plasticizers are di (2-ethylhexyl) phthalate (DEHP), di-isononyl phthalate (DINP) and di-isodecyl phthalate (DIDP). The largest single product used as a general purpose plasticizer worldwide is DEHP and it has set the standard for performance to price relationships since the 1950s. Several phthalates including DEHP are known as primary plasticizers due to their documented high compatibility with PVC polymers (Titow, 1984). However, the effects on health of DEHP, particularly with respect to asthma and allergies in children, has been a subject extensively investigated since the 1980s. Since January 2007, sale of toys and accessories intended for children younger than 3 years containing phthalates has been illegal in Europe (European Parliament Council, 2005).

Aliphatic esters are diesters of adipic, sebacic or azelaic acids and offer greater resistance to low temperatures than phthalates, but at a higher cost. Epoxy ester plasticizers have limited compatibility with PVC so are used at low concentrations. The most widely used epoxy ester plasticizer is epoxidized soya bean oil (ESBO). ESBO resists extraction, does not migrate readily and acts also as a heat stabilizer. Other plasticizers in this group include epoxidized linseed oil and tall oil.

Trimellitates are the esters of trimellitic anhydride (1,2,4-benzenetricarboxylic acid anhydride) and are noted for their low volatility. The most frequently used are trioctyl trimellitate (TOTM) and tri-isononyl trimellitate (TINTM). They are included in plastics which have to function at high temperatures for long periods and for PVC cable insulation in combination with phthalates. Phosphorous oxychloride reacts with various aliphatic and aromatic alcohols and phenols (triphenylphosphate) to produce triesters. Tricresyl phosphate was patented as a plasticizer for PVC in 1933, but was later found to be highly toxic and replaced. In addition to their role as plasticizers, phosphate esters, particularly triphenyl phosphate, function as flame retardants.

Fillers

Fillers are relatively unreactive, solid materials added to plastics formulations to modify their flow properties and handling during processing, as well as their tensile and compressive strengths, abrasion resistance, toughness, dimensional and thermal stabilities. They include naturally occurring organic materials such as finely divided wood, cellulose powder, starch, nutshells and protein fibres, inorganic compounds such as calcium carbonate and barium sulphate, glass beads, glass fibres and metals. Fillers are relatively inexpensive and can be used to replace some of the expensive pigment.

Fillers may be divided into particulate and fibrous types. Particulates include calcium carbonate, china clay, talc and barium sulphate. Fillers affect shrinkage on moulding and the dimensional stability of the finished plastic, increase tensile strength and hardness, enhance electrical insulation properties and reduce tackiness. They also impart opacity and colour (Figure 3.16). Carbon black is now the most widely used filler for polymers usually in the form of furnace carbon black, which has a particle diameter of 0.08 mm. Fibrous fillers reinforce polymers and greatly increase their tensile strengths. They include fibres of glass, textile and carbon. Plastics filled with fibrous fillers are known as composites.

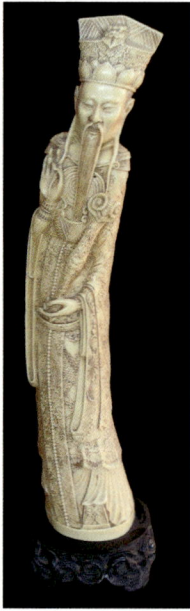

Figure 3.16

The polyester statuette is filled with calcium carbonate, which lowers the density of the piece, increases the opacity and imparts colour. The statuette resembles carved ivory, is 50 cm high and dates from the 1970s.

Particle size, particle shape, porosity, chemical properties of surfaces and impurities present in fillers influence the properties and stability of the final product. Particulate fillers range in diameter from 10 nm to macroscopic. The smaller the particles, the higher tensile strength, hardness and surface gloss of the filled plastic. Plastics filled with coarse particles tend to be weaker than unfilled materials. Particle shape affects tensile strength. Plate-like china clay fillers tend to be oriented during processing, which imparts a very high tensile strength in one direction. Unevenly shaped particles are more difficult to wet with liquid polymer than symmetrical ones. Traces of copper, manganese and iron present as impurities in fillers may accelerate oxidation in the finished plastic, resulting in weakening and discolouration.

Contact between filler particles and polymer is a key factor in the cohesiveness and properties of a plastic. If the ability of the polymer to wet or come in close contact with surfaces of the filler is poor, the plastic is likely to fail before its design limits have been reached. The chemical properties of fillers' surfaces influence how compatible they are with the liquid polymer. Polar groups at surfaces such as hydroxyls are more attracted to water than the polymer. To improve their attractiveness to polymer molecules, fillers are often treated with coupling agents. Stearic acid is used as a coupling agent for calcium carbonate. The polar acid groups in stearic acid attach to the filler particles while the aliphatic chain attaches to the polymer. Gycols are effective surface treatments for clays.

Wood flour or cellulose powder, both made by finely dividing wood waste material, have been included in phenol-formaldehyde formulations (Bakelite) since the early 1900s to minimize shrinkage on moulding and to increase compressive strength. Wood flour is clearly visible in dark plastics as white flecks (Figures 3.17 and 3.18). Finely divided cellulose is more fibrous than wood flour and is used as a filler for urea- and melamine-formaldehyde polymers. It has been proposed that the formaldehyde groups in the polymer react with the hydroxyl groups in cellulose (Carraher, 2000). Starch from plant materials has been used in quantities up to 30 per cent to produce biodegradable carrier bags from polyethylene. When buried in soil after use, bacteria obtain energy from the starch and degrade the polyethylene (The Guardian, 2003).

Figure 3.17
Wood flour used to fill phenol-formaldehyde is visible as pale flecks in the base of a caster or furniture cup.

Figure 3.18

Upper side of phenol-formaldehyde caster or furniture cup. White wood flour filler can be clearly seen in the small damaged area top right of figure.

Some fillers impart specialist secondary functions to the polymer. Glass in the form of fibres, flakes and hollow microballoons or microspheres has been used to fill polymers (Walker and Shashoua, 1996). Metal flakes or powdered metals are used as fillers in electrically conductive plastics, which are used to overcome electromagnetic interference in office equipment. Carbon-filled polymers are good conductors of electricity and heat.

Fibrous fillers employed in composites today comprise glass, carbon and graphite or aromatic nylons, also known as aramids. Asbestos fibres containing naturally occurring magnesium silicate have been used to impart heat and flame resistance to polymers, but their toxicity has greatly reduced the number of applications and asbestos now occupies less than 1 per cent of the fibre-composite market (Carraher, 2000). Reinforcing fibres require both higher tensile strength and stiffness than the polymer in which they are embedded so that the fibres bear most of the stress during use. Most fibres are thin, less than 20 μm thick, approximately one tenth the thickness of a human hair.

Glass fibres were first produced commercially in the mid 1930s and glass-reinforced polymer compositions were introduced in 1940. Glass fibre may be used in whole form, as woven cloth, chopped or stitched into matting. Various types of glass are available, including one which imparts good weathering resistance (low-alkali aluminium borosilicate) and another which can withstand high pressure (magnesium aluminium silicate). Glass fibres are pulled from melted glass, producing fibres 2–25 μm in diameter. Pulling orients fibres, increasing their strength and stiffness in the direction of the pull. Surface imperfections reduce strength and adhesion of polymer to the fibres. Glass fibres are impregnated with liquid polymer or powdered polymer before curing, hardening or heating into bulk forms and sheets known as glass-reinforced polyester (GRP) or fibreglass.

Close contact between filler fibres and polymer is essential to ensure good cohesiveness and physical stability of the final plastic. Attractive forces between fibrous glass and polymers are so weak that coupling agents are used to link them via covalent bonds. Agents containing silanes are the most widely used. They join to glass fibres via siloxane linkages (Si—O—Si) and to the polymer via an unsaturated chemical group (Figure 3.19).

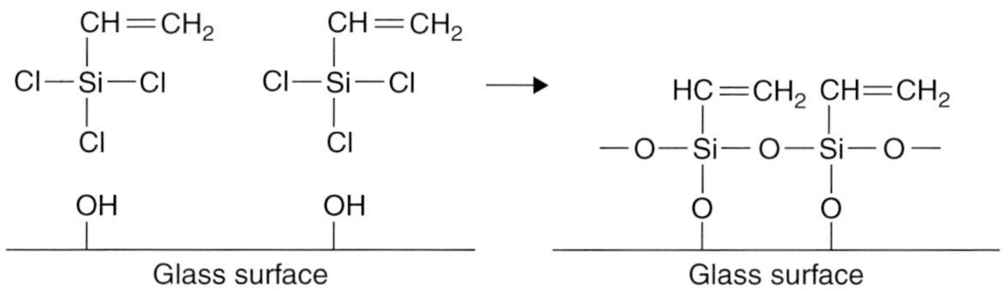

Figure 3.19

Silane-based coupling agents are most often used to link glass-reinforcing fibres to polymers. Siloxane linkages (Si—O—Si) are formed with hydroxyl (OH) groups of glass (to right of figure).

Unsaturated polyesters (UP), epoxies and formaldehyde-containing polymers are those most frequently reinforced with fibrous fillers. The polymer component is often referred to as the matrix. The advantages of reinforcing UP, particularly with respect to tensile strength and stiffness (E-modulus) properties, can be clearly seen in Table 3.5. Their low densities allow reinforced polymers to compete with traditional construction materials such as wood and steel (Aurer and Kasper, 2003).

Most boats, from small rowing boats to seagoing ships, are composed of GRP, utilizing its high tensile strength and its resistance to attack by both fresh and salt water. The low weight and resistance to corrosion of GRP compared with metals has led to extensive use of composites in the automotive and aerospace industries. More than $1000\,m^2$ of GRP is present in a Boeing 747.

Carbon and graphite fibres are produced by pyrolyzing synthetic fibres which align and form sheets comprising 93–95 per cent carbon. Carbon is also used in the form of whiskers which are sheets of hexagonal carbon atoms arranged in layers. In general, carbon fibres provide very high tensile strength but low interlaminar strength. The pyrolysis temperature controls the final properties of fibres. Carbon heated to 1500–2000°C produces very high tensile strength (5650 MPa), while carbon heated to 2500–3000°C imparts a high modulus of elasticity or stiffness (531 GPa). Carbon fibres may also be woven into cloth prior to impregnation with polymer.

	Non-reinforced UP	Glass-reinforced UP with chopped mat	Glass-reinforced UP with unidirectional fibres	Construction steel	Aluminium	Wood
Table 3.5 Mechanical properties of composites compared with steel, aluminium and wood						
glass content (% by weight)	0	30–35	75	0	0	0
tensile strength (MPa)	75	120	1200	390	250	100–180
E-modulus (MPa)	3300	8500	40 000	210 000	80 000	8000–15 000
density (kg/m³)	1300	1450	2000	7800	2700	600–1100
strength per weight unit (MPa/kg)	1.1	1.7	12	1	1.87	3.3

The low density of carbon fibre composites compared with metals has resulted in their application as biomaterials. Carbon fibre/epoxy composites are used as plates in bone surgery, replacing the titanium plates previously employed. The combination of high strength and flexibility exhibited by carbon fibre composites has had a huge impact in the performance of sports equipment. Fishing rods, golf clubs, bicycle frames, rackets and skis are examples of applications for carbon fibre composites.

Lubricants and mould release agents

Lubricants improve the flow characteristics of a material during processing. They function either by reducing melt viscosity of the polymer or by decreasing adhesion between the metal walls of processing equipment and hot liquid polymer. External lubricants exude during processing from the polymer to the metal surfaces of the processing equipment. They are poorly compatible with the polymer and often have a greater affinity to metals. The resulting thin film helps to prevent adhesion of the melted polymer to the equipment. Typical lubricants include stearic acid, calcium, lead, cadmium and barium salts and paraffin wax (Figures 3.20 and 3.21). External lubricants can be solid at ambient temperature but must be liquid under processing conditions. Because they exude from the plastic mass under processing and may dissolve other components from the formulation, lubricants should not be present at greater concentrations than 1 per cent (Brydson, 1999).

Internal lubricants are low molecular weight materials which promote flow of the polymer in the melt. Unlike external lubricants, compatibility between lubricant and polymer is essential. Waxes and stearates are typically used.

Impact modifiers

Impact modifiers are added primarily to PVC, polyethylene, polypropylene, polyamides and polyesters. They absorb the energy generated by impact and dissipate it in a non-destructive way. Impact modifiers are physically rubbery and semi-compatible with the polymer. The mechanism of absorbing impact in polymers is not fully understood, but these additives increase the tensile strength of the material. The impact modifiers most commonly used include acrylonitrile-butadiene-styrene (ABS) polymers, acrylics and ethylene-vinyl acetate (Carraher, 2000). ABS generates opacity or stress-whitening when used as

Figure 3.21
Stearic acid lubricant is
visible as a coating on a
PVC doll's legs.

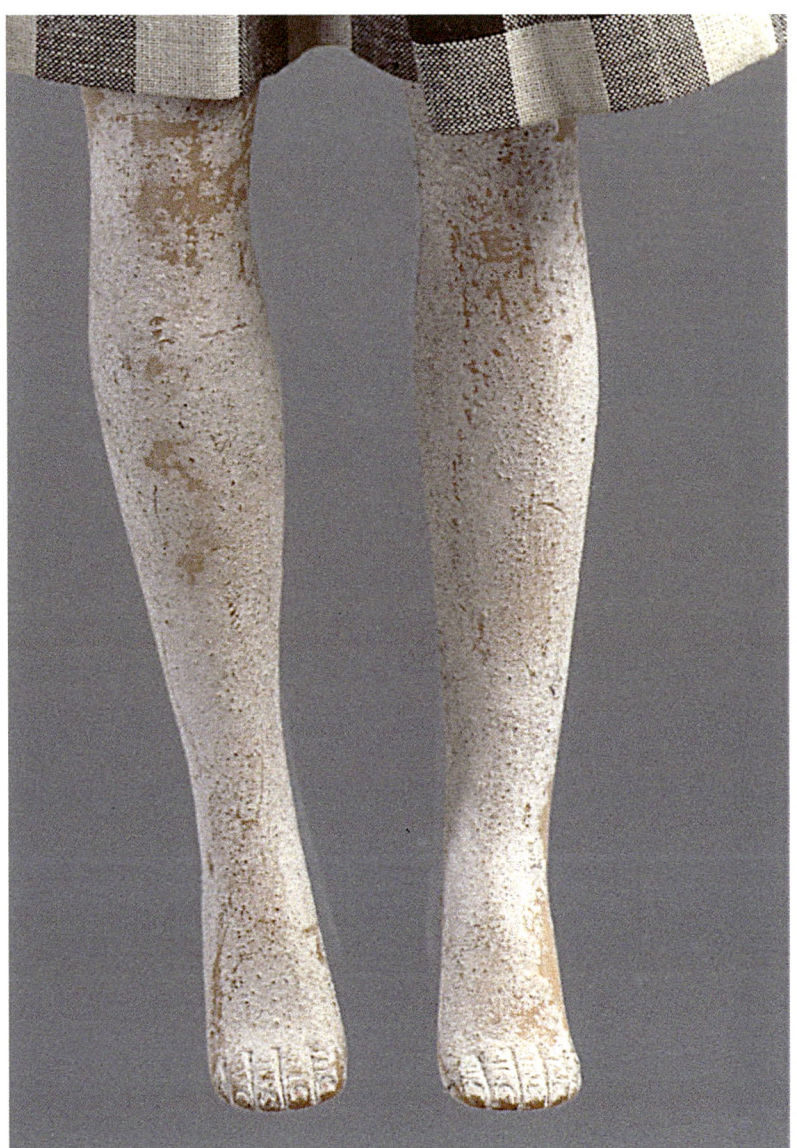

an impact modifier. This can be illustrated by PVC labelling systems, such as
Dymo® plastic tapes, which use the effect to form white, raised lettering on
receiving impact from a form (Figure 3.22).

Blowing agents

The application of a blowing agent is one method of producing a foam structure, that is, one in which a polymer matrix contains gas-filled cells. Chemically
foamed rubbers and plastics were first introduced in the late 1930s. Foams con-
sist of either discrete unit cells (unicellular) or interconnecting cells (multicellular) depending on the viscosity of the polymer. Unicellular foams are used in

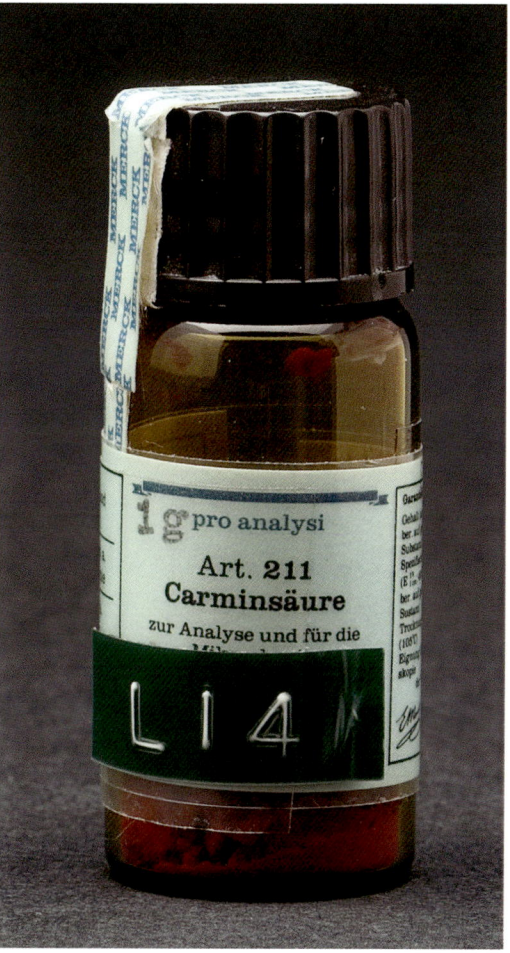

Figure 3.22
The impact modifier in PVC Dymo® tapes produces the white lettering.

applications which require insulation and buoyancy while multicellular foams are used in upholstery, carpet backing and laminated textiles (Figure 3.23). Cell size is influenced by the pressure of the gas and efficiency of its dispersion.

Blowing agents may be divided into physical and chemical types. Physical blowing agents are volatile liquids and gases which are first dissolved in the polymer, then evaporate (in the case of liquids) and expand on increase in pressure or temperature during processing. They comprise around 90 per cent of the world market (Harper, 2000). Liquid blowing agents comprise solvents with low boiling points such as aliphatic, fluorinated and chlorinated hydrocarbons. The most commonly used gases are carbon dioxide, nitrogen and air. Permeability within the polymer and the volume of gas released per unit weight of agent are measures of the efficiency of blowing agents. A volume of at least 150–200 mL of gas should be released per gram of agent under ambient conditions for the blowing agent to be effective. Physical blowing agents are particularly important in the production of polystyrene and polyurethane foams.

Figure 3.23
This family of German 'hedgehog dolls', dating from the 1960s, is constructed from unicellular polyurethane foam which allows them to be flexible while keeping their shape.

Chemical blowing agents decompose at high temperatures to evolve gas which expands the polymer matrix to produce a foam. Gas formation must take place close to processing temperatures for the polymer matrix and at a controlled rate. Exothermic blowing agents generate heat during decomposition and plastics foamed with these agents must be cooled for long periods to avoid thermal breakdown. They produce primarily nitrogen. Nitrogen is a more efficient expanding gas than carbon dioxide due to its slower rate of diffusion through polymers (Harper, 2000). Endothermic blowing agents require energy to decompose and produce primarily carbon dioxide. The most widely used exothermic chemical blowing agent for plastic foams, especially PVC and polyolefins, is azocarbonamide (chemical structure $NH_2CON{=}NHCONH_2$), a yellow-orange powder which decomposes at 190–230°C to produce nitrogen, carbon monoxide and dioxide.

3.2.2 Additives which affect chemical properties of polymers

Additives which change chemical properties during processing and in the final plastic product include anti-ageing additives in the form of antioxidants, heat stabilizers, light stabilizers and flame retardants.

Antioxidants
Antioxidants are added to polymers to retard oxidation. Oxidation is initiated by highly reactive free radicals which are formed by the action of heat, ultraviolet radiation, mechanical action and metallic impurities during polymerization, processing or use. The free radical can then react with an oxygen molecule to produce a peroxy radical (ROO*), a process known as propagation. The peroxy reacts with a hydrogen atom in the polymer to form an unstable hydroperoxide (ROOH) and another free radical (Feller, 2002). If these processes are allowed to continue, the polymer oxidizes. Oxidation causes crosslinking or chain scission.

Crosslinking results in an increase in molecular weight, brittleness and stiffness. Chain scission results in reduced molecular weight, softness and loss in tensile strength.

Antioxidants act to prevent the propagation stage of oxidation, and they achieve this by reacting with the free radicals formed during oxidation (Figure 3.24). Primary antioxidants donate their reactive hydrogen to the peroxy radical so that the propagation of subsequent free radicals does not occur, thereby quenching or stopping the chain reaction. The most frequently used primary antioxidants are sterically hindered phenols, but amines are highly effective and metal salts are also used. Secondary antioxidants decompose free radicals to more stable products. Secondary antioxidants are typically used in combination with primary antioxidants. Organophosphites and thioesters are the most frequently used secondary antioxidants.

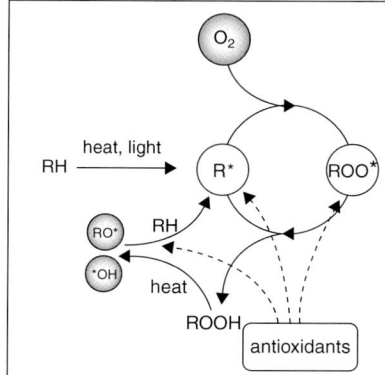

Figure 3.24

Antioxidants react with the free radicals (including R* and ROO*) formed during attack by oxidation of polymers (RH) in the presence of light, heat and metals. Antioxidants stop the free radicals from reacting further, thereby protecting polymers from deterioration.

Arylamines act as primary antioxidants by donating hydrogen. They are highly effective because they act both as chain terminators and peroxide decomposers. Negative properties include their tendency to discolour and stain and their toxicity, which precludes their application in non-pigmented plastics and those intended for food use. Arylamines are included in black wire and cable insulation and polyurethanes.

Phenolics are the most widely used antioxidants in plastics and are added to polyethylene and polystyrene and its copolymers. They may form coloured degradation quinine products on oxidation. The most widely used phenolic is butylated hydroxytoluene (BHT), which has an IUPAC name 2,6-di-*t*-butyl-4-methylphenol. BHT is added to many polymers including those used for food use, although its high volatility makes it unsuitable for plastics with long active lifetimes. Today, polyphenolics are replacing simple phenolics due to their higher molecular weight and resulting reduced volatility, although they are more expensive. A well-known brand name for polyphenolics is Irganox 1010.

Organophosphites are secondary antioxidants which reduce hydroperoxides to alcohols. They inhibit the discolouration reaction experienced by phenolics. Tris-nonyl phenyl phosphite (TNPP) is the most commonly used. The disadvantage of phosphates is their high hygroscopicity. Thioesters act as secondary antioxidants by destroying hydroperoxides to form stable sulphur derivatives. In addition, thioesters impart high heat stability to polyolefins, polystyrene and its copolymers. The major disadvantage of thioesters is their unpleasant odour which is transferred to the host polymer.

Metal deactivators are used in situations where plastics come into contact with metals such as polyethylene cable covers. These antioxidants combine with metal ions to form a stable complex at the interface, thereby preventing metals from accelerating degradation of the polymer.

Light stabilizers
Ultraviolet radiation (UV) with wavelength range 280–400 nm contains sufficient energy to cleave covalent bonds causing yellowing, surface cracking, hardening and changes in electrical properties of polymers, particularly if oxygen is also present. Polyethylene, PVC, polystyrene, polyesters and polypropylene are degraded by radiation at wavelengths 300, 310, 319, 325 and 370 nm respectively (Carraher, 2000). Light stabilizers protect polymers from degradation by light via three pathways: screening UV, absorbing UV and quenching or stopping deterioration reactions.

Light screens absorb radiation before it reaches polymer surfaces or reduce light's penetration into the plastic (Figure 3.25). Pigments in surface coatings act as light screens by absorbing radiation. Dispersing finely divided carbon black particles (15–25 nm diameter) into polymers at a concentration of at least 2 per cent by weight is highly effective.

Ultraviolet absorbers absorb primarily in the UV range while avoiding absorption in the visible light range so that the appearance of the plastic material's colour is not changed. Once absorbed, UV radiation may be converted to heat or to new chemical products. The most important commercial absorbers include o-hydroxybenzophenones, o- hydroxyphenylbenzotriazoles and salicylates. These convert absorbed radiation to heat.

Quenching agents do not absorb ultraviolet radiation but stabilize polymers by reacting with the free radicals generated by degrading polymers to stop the chain reaction. Hindered amine light stabilizers (HALS) were introduced in 1975 and are the newest of the light stabilizers. They have replaced benzophenones and benzotriazoles in polyolefins and are more cost-effective.

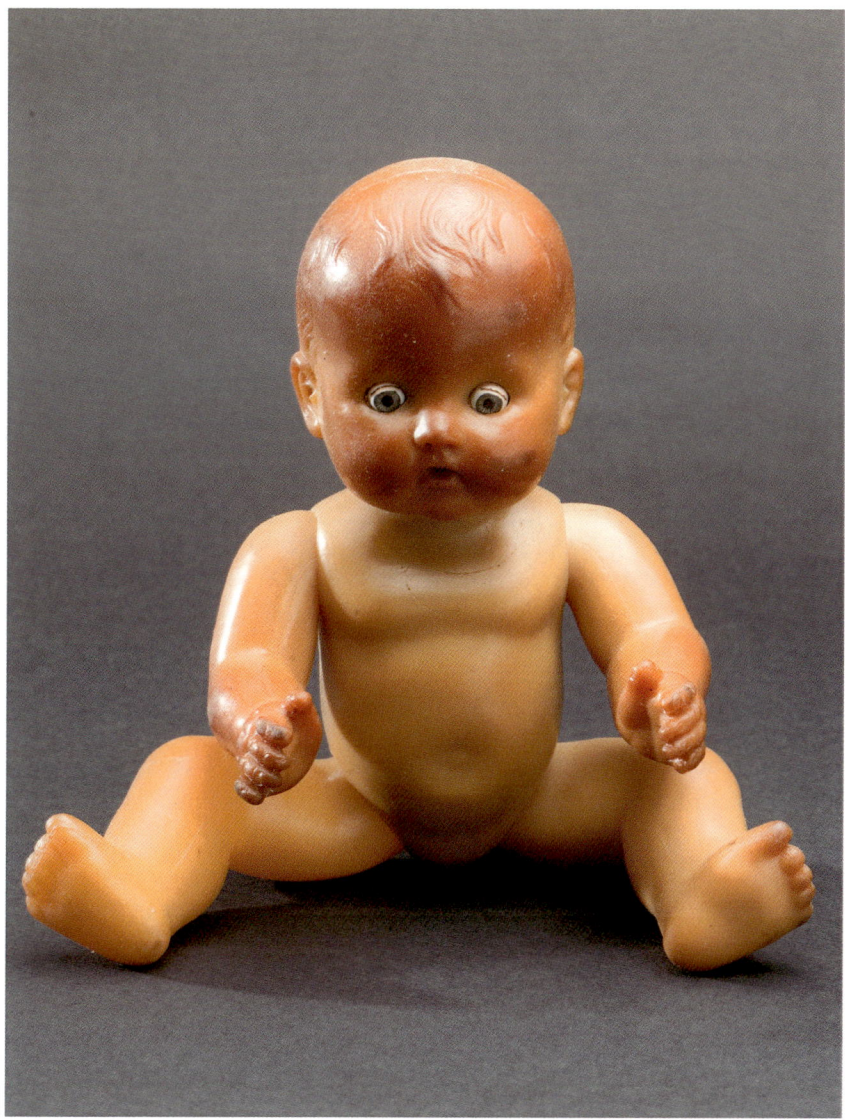

Figure 3.25
This PVC doll from
the 1950s has been
exposed to ultraviolet
light, causing severe
discolouration to her
face, lower legs and
lower arms. Her body,
upper arms and upper
legs are not discoloured
because she had a dress
which absorbed the
UV radiation before
it reached the PVC
surfaces. The dress
acted as a light screen.

HALS also act as antioxidants. Nickel (II) chelates act as quenching agents in polypropylene (Brydson, 1999).

Heat stabilizers

Heat stabilizers are added to protect polymers against thermal degradation during periods of exposure to elevated temperatures, including shaping by melt extrusion and during use at high temperatures. Most heat stabilizers are used to stabilize chlorine-containing polymers and copolymers, including PVC and vinyl chloride/vinyl acetate copolymers used to produce LP vinyl records. These materials are vulnerable because they produce hydrogen chloride gas (dehydrochlorination) on heating which discolours and darkens.

Heat stabilizers may be classified as either primary or secondary. Primary heat stabilizers inhibit dehydrochlorination in chlorine-containing polymers and react with any liberated hydrogen chloride to delay further degradation. Mixed metal salts are primary heat stabilizers and form metal chlorides with hydrogen chloride. However, mixed metal salts have a destabilizing effect which results in discolouration of the polymer. This effect is counteracted by the introduction of secondary heat stabilizers such as organophosphites and epoxy compounds. Organotin and lead primary heat stabilizers can be used alone.

The most common mixed metal stabilizers are octoates, naphthenates and benzoates of barium and zinc metals which are liquids. Barium, cadmium and zinc stabilizers contain solid salts of fatty acids including stearates and laureates. Barium and cadmium provide the most effective thermal stability followed by barium and zinc and then calcium and zinc. However, cadmium is currently considered undesirable from environmental and health standpoints and is being replaced in heat stabilizers, particularly those for polymers used with food.

Lead heat stabilizers are used for wire and cable applications as primary additives. They may be based on organic compounds including stearates and phthalates or inorganic salts such as sulphates, phosphates and carbonates. Although knowledge of the toxicity of lead has initiated the search for its replacement, no suitable, cost-effective material has yet been identified.

Alkyl and aryl organophosphites are often used as secondary heat stabilizers in conjunction with primary mixed metal salts. They prevent discolouration by forming complexes with the products of reaction between primary heat stabilizers and hydrogen chloride, such as barium chloride. Typical organophosphites include didecylphenyl, and triphenyl phosphites. Epoxidized soya bean and linseed oils function both as heat stabilizers and plasticizers in chlorine-containing polymers. They scavenge liberated hydrogen chloride and react with the polymer chain to prevent mobility of chlorine (Figure 3.26).

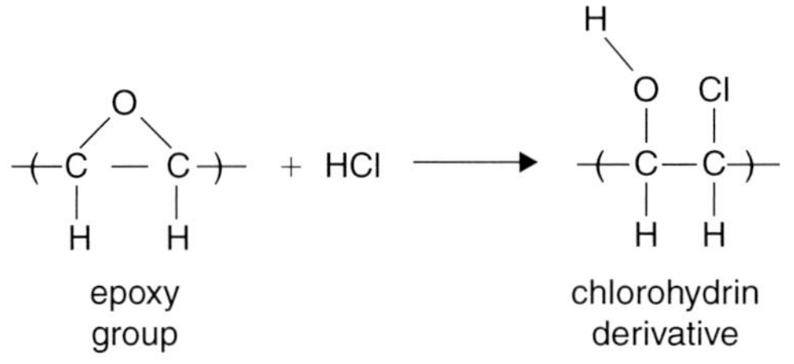

Figure 3.26

Epoxidized oils function as heat stabilizers in chlorine-containing polymers by scavenging liberated hydrogen chloride.

Flame retardants

Most polymers burn readily at elevated temperatures or support combustion when ignited with a flame. Halogenated polymers such as PVC show more resistance than non-halogenated materials. Some plastics, particularly polyurethane and PVC, produce toxic gases when ignited, such as carbon monoxide, hydrogen chloride and hydrogen cyanide. Because plastics are used in buildings, transportation, electrical appliances, clothing, furnishing and many other applications where people and animals come in contact with them, flame retardants are essential additives. Today, there are hundreds of different flame retardants available with various properties and costs. Some chemically bond with the polymer while others are simply mixed physically with them. They may be added at levels from 2 per cent to around 60 per cent of the total weight of a formulation, so are likely to affect the physical properties of the resulting plastic. The subject of combustion is a complex one and details can be found in specialized texts (Beyler and Hirschler, 2002).

Flame retardants work by minimizing at least one of the three factors necessary for fire to exist, i.e. fuel, heat and oxygen. They may produce char from the polymer which shields the surfaces from oxygen, retards the diffusion of volatile combustible products and insulates plastic from the heat of fire. Phosphorus and boron flame retardants catalyse the formation of char.

Metal hydrates such as aluminium trihydrate or magnesium hydroxide remove heat by using it to evaporate water in their structures, thus protecting polymers. Bromine or chlorine-containing fire retardants interfere with the reactions in flames and quench them. Mixtures of flame retardants antimony trioxide and organic bromine compounds are more effective at slowing the rate of burning than the individual flame retardants alone.

3.2.3   Additives which affect the appearance of polymers

Materials which are added to impart colour to polymers include soluble dyes and finely divided pigments. However, other additives also contribute to the appearance of plastics, particularly fillers calcium carbonate and carbon black.

Pigments are classified as either inorganic or organic. Inorganic pigments give opacity and have lower density and larger particles than organic pigments. Titanium dioxide is the most widely used inorganic pigment, imparting a white colouration. Titanium dioxide particles offer protection from ultraviolet radiation by reflecting light. The concentration added varies dramatically with location. Unplasticized PVC can be protected from UV radiation in temperate climates with around 4 per cent by weight titanium dioxide compared with 8 per cent if the plastic is exposed to tropical climates. Iron oxides or ochres produce yellow, red, black, brown and tan colouration and are the second most widely used pigments. Other inorganic pigments included in plastics formulations include yellow lead chromate, molybdate orange, yellow cadmium and green zinc chromate, but their high toxicities limit their applications. Ultramarine blue is widely used as a pigment in plastics.

Carbon black is the most widely used organic pigment closely followed by phthalocyanine blues and greens. Azo dyes include pyrazlone reds, diarylide yellows, dianisidine orange and tolyl orange. Quinacridone dyes include quinacridone violet, magenta and red. Other dyes used to colour plastics include anthraquinones such as flavanthrone yellow, dioxazines such as carbazole violet, and red and yellow isindolines (Carraher, 2000).

3.3   Shaping of plastics

The process of shaping plastics is known as converting because the process changes liquid, flowing plastic into a predetermined shape. Plastics may be converted into films or sheets, solid or hollow three-dimensional forms, fibres and foams. At the end of the nineteenth century, semi-synthetic plastics, like the natural materials they had been developed to mimic, were roughly moulded or shaped, then hand carved and polished. Hand finishing was a time- and skill-demanding activity. As demand for plastics grew, hand techniques were replaced by mechanized, automated bulk processes.

A variety of shaping processes is used to produce three-dimensional plastics today, many of which were developed before the 1940s in the USA and Germany (Brachert, 2002). Selection of appropriate processes depends on economic factors, the number, weight and dimensions of finished parts, the suitability of a particular plastic's thermal properties to a process, and the complexity of finishing operations. Mass production processes are frequently used to shape objects found in modern history, ethnographic, medical, sports history, furniture, military and children's collections.

It has been asserted that modern art does not use established, industrial processes and materials, nonetheless artists need material to realize even the most conceptual or metaphoric work (Beerkens, 2002). Although some artists state that they are not interested in using engineering processes, including sculptor Phillip King who works mainly with polystyrene and fibreglass, they often base their works on industrial materials (Coplans, 1965). Since few artists are also chemists, they select plastics materials from those produced in bulk for packing, wrapping, insulation and building, but may manipulate, adapt, reuse or misuse the materials and shaping processes to achieve their goal. Where industrial, bulk moulding processes are employed by artists, hand finishing techniques are generally used to customize pieces. An overview of commonly used techniques to shape both mass produced three-dimensional plastics and individual pieces is presented here. The production of fibres is not described because it is discussed comprehensively in another book aimed at museum professionals (Tímár-Balázsy & Eastop, 1998).

Calendering and extrusion are used to make plastic films. Three-dimensional solid plastics are formed by injection moulding, extrusion, blow moulding, thermoforming, rotational moulding, compression moulding, casting and encapsulation. Foamed plastic is shaped using injection, extrusion and compression moulding in combination with a foaming process. Reinforced plastics are shaped by hand lay-up techniques, by injection moulding, pultrusion or by transfering the desired form to prepared sheet or dough containing fibres. Selection of technique is based primarily on whether the material to be shaped is thermoplastic or thermosetting. Thermoplastics are vulnerable to distortion until cooled and this increases production time.

3.3.1   Shaping of plastic films and sheets

Most films are produced by extrusion. Thermoplastic materials are heated and pushed through a die to form a flat tube. Warm air is blown into the extruded tube to produce a balloon which is then cut open and laid flat. Calendering can also be employed to produce film. Hot thermoplastic is passed through a series of temperature-controlled metal rolls with progressively smaller gaps to produce a continuous sheet (Figures 3.27 and 3.28). The technique is used to make shower curtains, food wrap films, carrier bags and protective films. The most widely used films are polyethylene, nylons, polypropylene, cellulose acetate, PVC and polyesters.

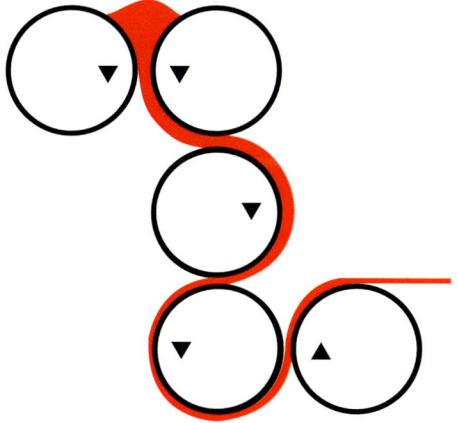

Figure 3.27
A calender is a set of heated, rotating rollers used to make films and sheets, with progressively smaller gap widths between each roller from the first to the last.

Figure 3.28
A poster made of PVC sheet by calendering.

3.3.2   Shaping of 3D solid plastics

Progress made in injection moulding since the 1920s has resulted in high speed production and flexibility in processing thermoplastics, thermosets and elastomers. Today, injection moulding is the most common technique for processing plastics. Plastic granules or pellets are heated in a cylinder until they have a sufficiently low viscosity to be injected under pressure into a cold, split mould (Figure 3.29). Moulds can be designed so that no seam is visible on the finished product. Heating thermoplastics requires care if overheating, which can result in degradation, is to be avoided. A screw or a piston pushes the solid pellets from a hopper towards a heated cylinder where the plastic is melted. The plastic is then forced into the water-cooled mould.

All plastics contract on cooling and large differences in thickness between sections of an object will encourage the formation of sink marks or voids, which can cause distortion, warping or collapse. The plastic component is ejected when the mould opens and the cycle is repeated, each cycle lasting 3–5 minutes depending on the thickness of the plastic. The rate of injection moulding is increased by cooling the mould so that thermoplastics harden before being ejected. Plastic objects shaped by injection moulding include CDs, DVDs, vehicle components, beer crates, telephones and buckets (Figure 3.30).

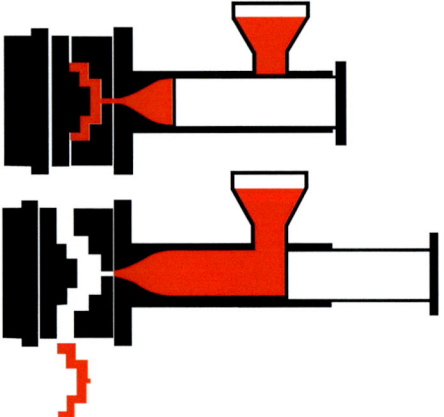

Figure 3.29

In injection moulding, plastic granules are heated until they have a sufficiently low viscosity to be injected under pressure into a cold, split mould. The shaped plastic is cooled and released.

Figure 3.30

The polycarbonate base in a CD is injection moulded.

Injection moulding can be applied to two-component thermosets in a process known as Reaction Injection Moulding. The two liquid components are mixed and then injected into a closed mould. In the case of epoxies and polyurethanes, heat is produced on mixing due to the exothermic reaction between the two components (polyol and isocyanates in the case of polyurethanes), so no external source of heat is required.

Extrusion moulding was first developed in the 1830s. In the extrusion process, thermoplastic granules or pellets are heated to melting. Extrusion takes place under high pressure through a shaped die to produce symmetrical, continous forms with even thicknesses such as pipes, fibres and films (Figure 3.31). Drain pipes, guttering, plastic window frames, roofing sheets and many other products are formed by extrusion. Extrusion moulding has a very high output rate. For example, pipe can be produced at a rate of 900 kg/h (US Environmental Protection Agency, 2005).

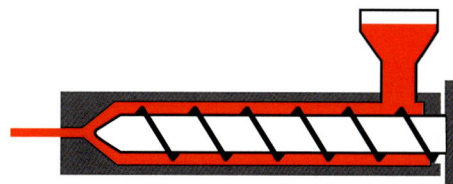

Figure 3.31

In the extrusion process, plastic granules are introduced through a funnel (right-hand side of diagram) and heated to melting. Liquid plastic is transported and pressed by a rotating screw through a shaped die. A continuous shape of even thickness is formed (left-hand side of diagram).

In 2002, designer Tom Dixon designed and made a collection of chairs, bowls and coffee tables from extruded and hand-woven polyester called Fresh Fat. His intention was to apply a handicraft similar to weaving or basket making to the industrial extrusion process. Eastman Provista®, PETG polyester, was extruded in 5-mm-diameter tubes and folded on itself in loops while still warm to build in layers the required shape of the bowl, chair or table (Figure 3.32). On cooling, the polyester was glass clear and brittle. Tom Dixon said 'we are going in exactly the opposite direction to the whole history of plastic manufacturing which has continually driven towards consistency. With this process it is impossible to make two objects that are exactly the same' (Dixon, 2002).

Blow moulding describes any shaping process in which air is used to stretch and form plastic materials and can be compared to inflating a balloon. A hot thermoplastic tube, usually made by extrusion, can be inflated with compressed gas while inside a cooled split mould. Hot thermoplastic tubes or parisons can also be blown into free shapes without the aid of a mould (Figure 3.33). Plastic bottles, drums, car fuel tanks and other containers are often made using blow moulding (Figure 3.34).

In the thermoforming process, heat and pressure are applied to thermoplastic sheets which are placed over moulds to adopt their shapes (Figure 3.35). Pressure can be applied using air, compression or vacuum (vacuum forming). Compression is a low cost process because the moulds used are inexpensive and automation allows a production cycle of less than 2 seconds. Details are

Figure 3.32

Tom Dixon's Fresh Fat furniture and bowls are made by folding extruded tubes of polyester while warm.

Figure 3.34

Blow moulded soft drinks bottles are made without seams from polyester.

Figure 3.33

In blow moulding, a pre-formed, thermoplastic tube is reheated and blown to shape in a cooled mould before release.

lost if thick sheets are thermoformed. Plastics in sheet form cost more than the plastic granules used in other moulding and shaping processes. Products of thermoforming include shaped packaging such as blister packaging, yoghurt pots and vending cups (Figure 3.36).

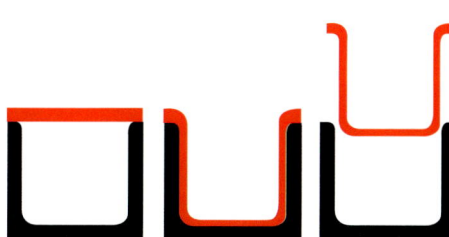

Figure 3.35
During thermoforming, a thermoplastic sheet is heated and either sucked or blown to adopt a form before cooling.

Figure 3.36
Products of thermoforming include food containers with complex shapes and undercuts.

In rotational moulding, thermoplastics are heated, shaped and cooled in the mould. Finely ground plastic powders are heated to melting and tumbled in a split, hollow mould. The inner surfaces of the mould are evenly coated with molten plastic due to centrifugal force and allowed to cool until hard. The final products are hollow with uniform thickness and relatively low moulding costs. Typical products include litter bins, traffic cones, fuel tanks and plastic balls.

The invention of Bakelite in 1907 was the drive behind the commercial use of compression moulding and resulted in the technique becoming the most common technique for the production of thermosetting plastic products today (Brachert, 2002). A predetermined mass of polymer in powder or pellet form is mixed with additives and reinforcing materials, if necessary, and introduced to a lubricated, hardened steel mould which has been machined to size. Pressure is applied to close the mould and causes the mixture to adopt the mould's dimensions (Figure 3.37). Application of heat hardens the plastic in its moulded form although allowances must be made for shrinkage. Selection of temperature depends on the plastic type with phenol-formaldehydes requiring 177°C while urea-formaldehydes require 149°C (Brachert, 2002). Bang & Olufsen developed their own Bakelite press in 1939 to mould the cabinet for the first Beolit radio (Figure 3.38) (Thoegersen, 2007).

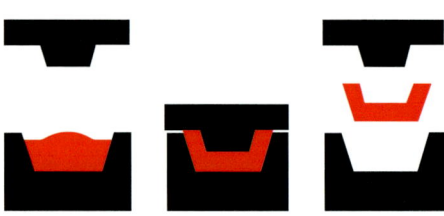

Figure 3.37

In compression moulding, solid polymer and additives are placed in the mould, pressure and heat are applied, and the product removed.

Figure 3.38

The Bakelite cabinet of the Beolit radio by Bang & Olufsen was compression moulded in 1939.

Wall thicknesses should be uniform in compression moulds because heating and cooling times are dependent on the mass of plastic present. Variation in temperature can result in uneven stiffness and cracking on removal from the mould. Where very thin walls or undercuts are present in an object, it is usually moulded in several sections which are adhered or slotted together. Fuse boxes, lamp holders and saucepan handles are produced by compression moulding. Compression moulding requires a longer cycle time than injection moulding which increases costs.

Intricate plastics products which contain deep holes or metal inserts are shaped via a similar method known as transfer moulding in which the plastic is liquefied prior to being injected into a closed mould. Fine details are preserved by starting with a flowing, liquid plastic instead of the powder used in compression moulding.

Dip moulding is a technique for making very thin walled, flexible products. Metal forms with identical dimensions to the final product are heated to the film forming temperature of the plastic being processed. The forms or mandrels are dipped rapidly into a bath of plastic paste or fluidized powder and withdrawn to allow excess material to drip. The cooled film is peeled from the forms to produce balloons and protective and medical gloves.

3.3.3   Shaping of 3D foamed plastics

Shaping of foams provides different challenges to shaping of liquid plastics because the foam grows and expands while being shaped. Foams may be described as either blown or structural types. Blown foam is used to describe a polymer matrix expanded by the introduction of air, carbon dioxide or nitrogen. Structural foam is a foamed core surrounded by a solid outer skin. It has a high strength to weight ratio and may have three to four times higher rigidity than a solid moulded part made from the same plastic type. Both types of foam can be shaped by injection, extrusion and compression moulding in the same way as solid plastics, but the blowing agent or other method by which air is introduced to the matrix participates in the moulding process (Figures 3.39 and 3.40).

Figure 3.39
Polystyrene foam plant pot shaped by injection moulding.

Blown foam is shaped by extruding a polymer matrix which contains a volatile liquid. The liquid evaporates when the polymer matrix is heated and produces cells in the plastic on cooling. This process is applied to polystyrene, PVC, polyethylene, urea-formaldehyde and acrylonitrile-butadiene-styrene polymers. Other techniques to introduce gases into the polymer matrix include dispersing air mechanically, which may be likened to whipping cream, injecting pressurized gas into the polymer matrix, and the introduction of a chemical blowing agent which generates nitrogen, carbon dioxide or carbon monoxide on heating. Polyurethane foams are produced by the reaction between a polyol and a diisocyanate. The polyurethane polymer ingredients are mixed and poured evenly into a moving trough or Henecke machine. Water and catalysts are then injected into the polymer and the whole is vigorously stirred. As foaming takes place, the mixture forms an even block of foam which, when cool and hard, may be cut to size. If a cylinder of foam is required, the foaming mixture is fed into the bottom of a

Figure 3.40
The point at which polystyrene was injected can be clearly seen as a raised area on the base.

cylinder and pushed upwards. Pressure applied above the foaming mixture is used to control density. The solidified cylinder may be sliced horizontally into discs.

Structural foam plastic is shaped by injection moulding a liquid polymer in which chemical blowing agents are dispersed. A lower volume of the mixture is injected into the mould than would be required for an equivalent solid. The initial injection pressure applied is so high that the chemical blowing agent cannot be activated and an outer, solid skin forms at the surfaces of the mould without foaming. After formation of an outer skin the injection pressure is reduced and the remaining polymer matrix expands to fill the rest of the mould as the chemical blowing agent is activated.

3.3.4   Shaping of reinforced plastics

Reinforced plastics first became commercially available in the 1940s. Unsaturated polyester reinforced with fibrous glass is the most widely used reinforced plastic today. Chopped glass mats are used for sheet moulding compounds (SMC), bulk moulding compounds (BMC) and hand lay-up composites.

Hand lay-up technique is the traditional process for shaping glass fibre-reinforced materials and is used today to make moulds, prototypes, boat hulls and decks. Release agent is first applied to the mould. A gelcoat containing colourants and additives is applied to the mould's surfaces and determines the visual surface of the finished object. The gelcoat also determines weather resistance, light fastness, gloss, and chemical and mechanical properties, though it is not reinforced with fibres. Gelcoats contain accelerators to initiate curing at room temperature, thickening agents to prevent sagging and are applied in thin layers (500–600 μm) by brush, roller or spray to minimize the inclusion of air bubbles.

Layers of glass fibre matting are placed over the gelcoat layer, if present, to line the mould. Glass fibres are coated with polyester matrix using rollers to remove air bubbles and ensure excellent wetting of fibres. After curing at room temperature, mouldings are removed and trimmed.

The Finnish artist, Kari Tykkylainen, used glass-reinforced polyester (GRP) to form the installation 'Cocotte with Two Dogs' in 1987. During an interview conducted by conservator Paivi Kyllonen at the Oulu City Art Museum in Finland, Tykkylainen explained that reinforced polyester was inexpensive and easy to use in the production of light, strong and rigid structures (Knuutinen and Kyllonen, 2006). The installation comprises a three-dimensional black female figure and two dark red-brown dogs with metal chains linking the three figures (Figure 3.41). The dogs were constructed by hand lay-up to produce hollow forms open on the front side (Figure 3.42). Glass fibres impregnated with polyester were laid directly over a clay form without first applying a gelcoat. The form was removed prior to the through hardening of the GRP. A specially made pigment mixture was applied directly to the reinforced polyester and transparent acrylic lacquer applied to pigmented areas to protect against ultraviolet radiation.

Reinforced plastics can also be shaped by injection moulding. Reinforcing material is cut to the shape of the final product and placed in the mould. Metal inserts such as screw threads for bolted portions and other materials required for the product to function are also added at this stage. The mould is closed and the polymer mixed with accelerator is injected with or without the application of vacuum.

Sheet moulding compound (SMC) is unpolymerized sheet comprising pre-impregnated sections of glass fibre which already contain the necessary accelerator and additives. SMC is coated on both sides with protective polyethylene or polyamide films, which allows it to be cut to size and handled safely prior to shaping. It is placed in a negative mould prior to curing using ultraviolet radiation or heat (Elias, 1997). Bulk moulding compound (BMC) is a dough-like material containing glass fibres, polymer, accelerator and additives. It can be injection or compression moulded to make high volume products such as car headlamp reflectors.

Figure 3.41

'*Cocotte with Two
Dogs*' (1987) by Kari
Tykkylainen.

3.3.5   *Finishing and decorating processes*

Shaping processes may result in rough mouldings which require finishing, polishing and decorating. If small leakages have occurred during shaping due to poor
fitting of multi-part moulds, incorrect temperature or pressure during moulding,
extra plastic can form along the separating line of the mould, known as flash.
Sprue describes rods and frames of plastic which connect several small plastic

Figure 3.42
The hollow dogs were constructed from glass fibre-reinforced polyester using hand lay-up. Red pigmented coating was applied.

components moulded in one piece, such as polystyrene airplane model kits. These unwanted additions are trimmed away or abraded to produce finished surfaces.

Machining is often necessary for plastics shaped by blow moulding and extrusion. Injection moulded parts normally do not require any machining operations, apart from removal of flash and sprue. Band saws and junior hacksaws can be used to remove additional plastic. High speed cutting and polishing tools used on metals are often applied to plastics, but the very low heat conductivity of plastics compared with metals can result in distortion and melting. Cutting oils and cooling liquids may not be chemically compatible with plastics and are replaced by air or water cooling.

Plastics can be drilled though they require clamping securely to avoid cracking. Polishing cut edges or surfaces of plastics is achieved using sandpaper or with commercial plastic polishes. Such products usually contain abrasive particles suspended in silicone polish or another liquid polymer. Mechanical fixing techniques including screws and rivets can be used to join plastic sections with care. The selection of an appropriate solvent-based adhesive to join sections is imperative because of the risk of dissolution or stress cracking of plastics. High-tech methods to make joins include ultrasonic welding.

Although plastic parts are usually produced with moulded-in colour, paint and coatings may be applied to hide irregularities, match colour or pattern with those of adjacent components or improve resistance to abrasion, chemicals or ultraviolet radiation. Painting with a coating that contains metal particles can give plastic parts a metal-like appearance and special electrically conductive paints can provide electromagnetic interference shielding. Selection of suitable paints and coatings is limited by knowledge of their contained solvents, which might dissolve or etch plastics. In addition, the difference between flexibility of the plastic component and the paint should be considered if cracking and loss of coatings during use is to be avoided. Powder coatings, often based on nylon and applied with heat, epoxy and polyurethane coatings, are those most commonly applied to plastics because they contain little or no solvent (DSM Engineering Plastics, 2006).

Hot transfers are used to provide parts with a decorative pattern or lettering. They comprise a pre-printed transfer film, on which the design has been applied. A heated spatula or plate is used to transfer coating from the film to the plastic. Water-based adhesives are also used to adhere printed transfers to plastics.

Peter Chang is a British artist who has specialized in making plastic jewellery since the 1970s (Figure 3.43). Peter Chang uses traditional tools to handmake

Figure 3.43
Artist Peter Chang's plastic bracelets involve many hours of hand finishing.

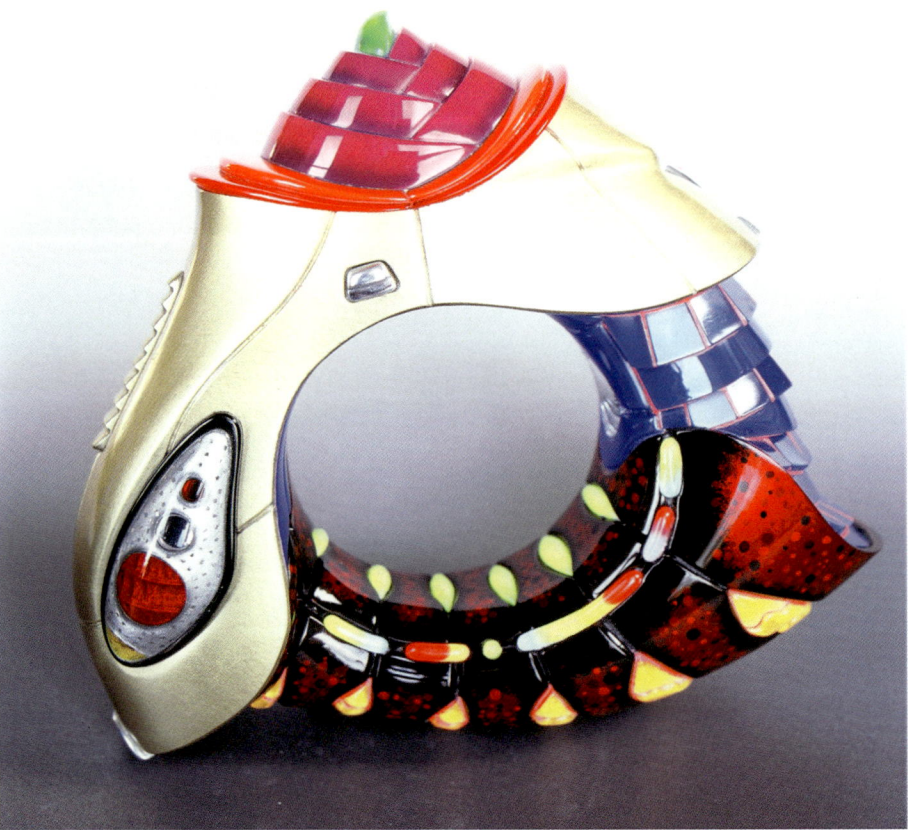

his jewellery including band saws, lathes, sanders, polishers and rotary drills. He has described the preparation of a bangle *Statement of Significance* (Holzach, 2002). The bangle's dimensions are 73 mm×215 mm×205 mm. It is highly coloured with stripes and spots, some raised. Projecting tentacles in stripes of laminated polyester and acrylic sheet in yellow, orange and red have been applied to the outside. The bangle was based on a commercially made polystyrene foam form which was hand carved to a doughnut shape. The form was decorated by applying layers of lacquer by brush and with carefully pos-itioned inlays of coloured acrylic and polyester sheet. Plastic beads and the polyethylene heads of dressmaker's pins were added as decorations. 'Statement of Significance' required around 246 hours to make, excluding drawing and design, and exemplifies the difference between commercial, bulk produced plastics and the creation of unique plastic pieces by artists and designers.

References

Aurer, J. H. and Kasper, A. (2003). *Unsaturated polyester resins, polymers with unlimited possibilities.* Verlag Moderne Industrie, Landsberg.

Beerkens, L. (2002). Plastics in modern art. In *Postprints of Plastics in Art, History, Technology, Preservation* Cologne, Spring 2001 (T. van Oosten, Y. Shashoua, F. Waentig, eds) pp. 7–17, Siegl, Munich.

Beyler, C. L. and Hirschler, M. M. (2002). Thermal decomposition of polymers. In *SFPE Handbook of Fire Protection Engineering*, 3rd edition (P. J. DiNemmo, ed.) pp. 1/110–1/131 NFPA, Quincy, MA, USA.

Brachert, E. (2002). Compression moulding and injection moulding – two ways of moulding plastics. In *Postprints of Plastics in Art, History, Technology, Preservation* Cologne, Spring 2001 (T. van Oosten, Y. Shashoua, F. Waentig, eds) pp. 34–44, Siegl, Munich.

Brydson, J. A. (1999). *Plastics Materials*, 6th edition. Butterworth-Heinemann, Oxford.

Carraher, C. E. Jr. (2000). *Polymer Chemistry*, 5th edition. Marcel Dekker Inc., New York.

Callister, W. D. Jr. (1994). *Materials science and engineering – an introduction*, 3rd edition. Wiley, New York.

Coplans, J. (1965). Phillip King. *Studio International*, 170, 254–257 [online]. Available from: http://www.studio-international.co.uk/archive/king-phillip-1965-170.asp [Accessed 27 January 2007].

DSM Engineering Plastics (2006). *Finishing and decoration* [online]. Available from: http://www.dsm.com/en_US/html/dep/finishing_decorations.html [Accessed 17 February 2007].

Dixon, T. (2002). *Tom Dixon* [online]. Available from: http://www.tomdixon.net/ [Accessed 20 February 2007].

Elias, H-G. (1997). *An introduction to polymer science.* VCH Publishers Inc., New York.

European Union Commission (2000). *Green paper on environmental issues of PVC (COM (2000) 469FINAL)* [online]. Available from: http://www.europa.eu.int/comm/environment/pvc/index.html [Accessed 24 February 2007].

European Parliament Council (2005). *Directive 2005/84/EC* [online]. Available from: http://eur-lex.europa.eu/LexUriServ/LexUriServ.do?uri=CELEX:32005L0084:EN:NOT [Accessed 24 February 2007].

Feller, R. L. (2002). Stages in the Deterioration of Organic Materials. In *Contributions to Conservation Science* (P. M. Whitmore, ed.) pp. 353–369. Carnegie Mellon University Press, Pittsburgh.

Harper, C. A. (ed.) (2000). *Modern Plastics Handbook*. McGraw-Hill, New York.

Holzach, C. (ed.) (2002). *Peter Chang: it's only plastic*. Arnoldsche Art Publishers, Stutgard.

Horie, C. V. (2002). *Materials for Conservation – organic consolidants, adhesives and coatings*. Butterworth-Heinemann, Oxford.

Kaufmann, M. (1963). *The First Century of Plastics – celluloid and its sequel*. Chameleon Press Ltd., London.

Knuutinen, U. and Kyllonen, P. (2006). Two case studies of unsaturated polyester composite art objects. *e-Preservation Science*, **3**, 11–19.

Mossman, S. (ed.) (1997). *Early Plastics Perspectives, 1850–1950*. Leicester University Press, London.

Quye, A. and Williamson, C. (1999). *Plastics, Collecting and Conserving*. NMS Publishing Ltd., Scotland.

Rodriguez, F. (1989). *Principles of Polymer Systems*. Hemisphere Publications, New York.

Tabb, D. L. and Koenig, J. L. (1975). Fourier Transform Infrared Study of Plasticized and Unplasticized Poly(vinyl chloride). *Macromolecules*, **8**(6), 929–934.

The Guardian. (2003). *New plastic bag will biodegrade in a month* [online]. Available from: http://www.smh.com.au/articles/2003/02/12/1044927664103.html. [Accessed 10 January 2007].

Thoegersen, J. (2007). *Bang and Olufsen* [online]. Available from: http://www.thogersen.dk [Accessed 20 February 2007].

Tímár-Balázsy, Á. and Eastop, D. (1998). *Chemical principles of textile conservation*. Butterworth-Heinemann, Oxford.

Titow, W. V. (1984). *PVC Technology*, 4th edition. Elsevier Applied Science, London.

US Environmental Protection Agency. (2005). *EPA Office of Compliance Sector Notebook Project, Profile of the Rubber and Plastic Industry*, 2nd edition. Chapters I, II and III [online]. Available from: http://www.epa.gov/compliance/resources/publications/assistance/sectors/notebooks/rubber.html [Accessed 20 February 2007].

Walker, W. and Shashoua, Y. (1996). Microballoon Fillers for Friable Ceramics. *Conservation News*, March, 11.

Wilson, A. S. (1995). *Plasticisers: principles and practice*. The Institute of Materials, London.

Properties of plastics

Summary

Chapter 4 presents an overview of the chemical, optical, physical and thermal properties of plastics which are most relevant to conservation. Properties determine which plastics are suited to particular functions and also why some are no longer in use. The chapter starts with a basic description of the types of bonding and structure which determine the properties of polymers. Details of the chemical, optical, physical and thermal properties for plastics most often encountered in collections are presented in tables, one for each material, in Appendix 1.

The properties of plastics may be divided into chemical, optical, physical and thermal. The aspects of each of those properties, which are of greatest relevance to conservation of plastics, are discussed here. The theory behind the major factors which determine each property are presented in this chapter. Details of the chemical, optical, physical and thermal properties for plastics most often encountered in collections are presented in tables, one for each material, in Appendix 1.

4.1 Chemical properties

The chemical properties of plastics of greatest interest to conservation professionals are their solubility and their reactivity with the materials in their surroundings. As a background to discussing solubility and chemical reactivity of plastics, it is important to be aware of the structure, arrangements and types of bonding present in polymers. A polymer is a large, complex molecule constructed from many smaller, simpler molecules. There may be one species of unit, repeatedly joined, or several different species which react together. Addition polymers comprise a long chain of one species. For example, polyethylene comprises between 1000 and 10000 methylene groups ($-CH_2-$) joined to form a chain as shown here.

$$-CH_2-CH_2-CH_2-CH_2-CH_2-CH_2-CH_2-CH_2-CH_2-$$

To synthesize polyethylene a double carbon bond in the starting material ethylene ($CH_2=CH_2$) breaks to allow attachment to other ethylene molecules resulting in a high molecular weight material or macromolecule. Other polymers which are formed by a similar process include polystyrene (repeat unit or monomer is styrene), polypropylene (monomer propene), poly (methyl methcrylate) where the monomer is methyl methacrylate, 1,4-polybutadiene (monomer is buta-1,3-diene) and 1,4-polyisoprene (monomer is isoprene) which has the same formula as natural rubber. Details of how polymers are prepared and processed are presented in Chapter 3.

Although all polymers have high molecular weights due to the development of long chains during polymerization, not all polymer chains polymerize to the same extent. As a result, most polymers comprise a distribution of chain lengths and molecular weights. The components with lower molecular weight will be more easily soluble and have lower tensile strength than those with higher molecular weight. The properties of polymers therefore represent the average for all components.

Box 4.1   Bonding in polymers

Covalent bonds are the most important forces which join atoms in polymers. They are formed by sharing one or more pairs of electrons between two atoms. Sharing is usually unequal, resulting in the formation of small negative and positive charges at the two ends of each bond, a phenomenon known as polarity. The greater the difference between the electronegativity values of the two constituent atoms, the greater the bond's polarity. The electronegativity value of carbon is 2.5, that of nitrogen and chlorine is 3.0, and of oxygen is 3.5 (Brydson, 1999). Thus a carbon-oxygen bond (difference 1.0) will be more polar than a carbon-chlorine bond (difference 0.5).

The energy required to destroy or dissociate covalent bonds largely determines the chemical and thermal stability of the polymer. A table and explanation of dissociation energies are presented in Chapter 6. Carbon single bonds are weaker than carbon-carbon double or triple bonds (348, 612 and 812 kJmol^{-1} respectively). Because many covalent bonds are polar, different areas within a polymer chain may carry opposite charges. Polarity leads to interaction between neighbouring chains through the formation of secondary, intermolecular forces. These forces may be defined as dipole, induction and dispersion. Hydrogen bonding is a type of secondary bonding formed when hydrogen atoms in hydroxyl (OH), amine (NH) or amide ($CONH_2$) groups are shared with a proton accepting group such as the oxygen in carbonyl ($C=O$), ether or hydroxyl, or the nitrogen atom in amines or amides. Secondary bonds have approximately one tenth the dissociation energy of covalent bonding, so are more readily destroyed.

4.1.1 Structure of polymers

Carbon atoms in all hydrocarbons, such as polyethylene, are joined at the characteristic tetrahedral bond angle of approximately 109.5°, have a bond length of 0.126 nm and can rotate and bend in three dimensions. When thousands of methylene groups are joined together at an angle greater than 90°, the carbon atoms produce a chain with a zigzag form which may twist, bend and rotate into other positions while maintaining the 109.5° angle (Figure 4.1). Although frequently described as straight chain or linear polymers, this is a simplification in the case of hydrocarbons. If there are many polymer chains present, each of which can bend and twist, this leads to extensive intertwining and entanglement between chains.

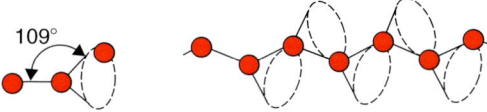

Figure 4.1
Polyethylene chains form a zigzag shape due to the characteristic tetrahedral bond angle of 109.5°. The chain can bend and rotate within the region marked by the dotted line.

Condensation polymers may be arranged in branched or network structures because the splitting off of a small molecule during synthesis creates more points at which monomers can join (Figure 4.2). Branches, considered to be part of the main chain of the polymer, result from side reactions which take place during synthesis. Branched polymers pack less efficiently than linear polymers. Joining of adjacent linear polymer chains, via crosslinks comprising covalent bonds, results in the formation of a three-dimensional network structure. Polymer chains which are connected by crosslinks cannot move freely past each other. Polymers may be either lightly or highly crosslinked – lightly crosslinked materials can deform and recover, such as synthetic rubbers, while highly crosslinked

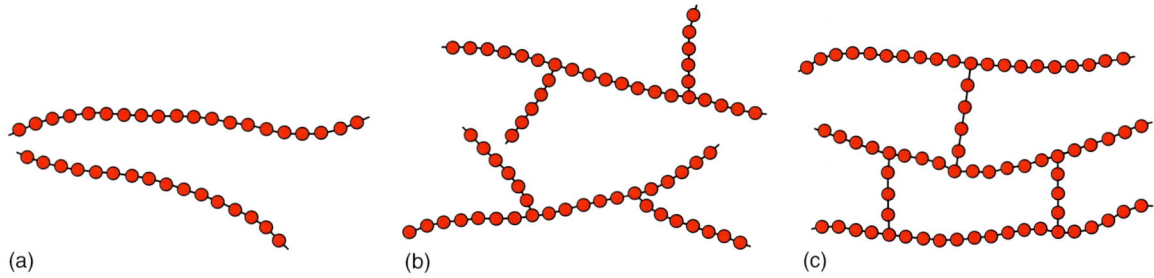

(a)   (b)   (c)

Figure 4.2
Polymer chain structures may be linear such as polyethylene (a), branched such as polyesters (b) or crosslinked such as epoxies (c).

polymers are more rigid and include epoxies and phenol-formaldehyde. In general, branched polymers have a higher molecular weight than linear ones while crosslinked structures have a considerably higher molecular weight than branched polymers. Polymers rarely comprise one structural type exclusively and a linear polymer may contain some branching and crosslinking.

Linear and slightly branched polymers with a regular structure can form crystals. Polymers with an irregular structure do not form crystals and are described as amorphous. In the early 1920s, Haworth was the first to identify the presence of crystalline areas in a polymer, employing X-ray diffraction to examine cellulose (Carraher, 2000). Crystallinity in inorganic materials differs from that in polymers. The former comprises a single crystal developed from a nucleus whereas polymers develop clusters of single crystals simultaneously from many nuclei, a property known as polycrystallinity found also in metals (Brydson, 1999).

Recent research suggests that polymer molecules or chains align themselves and then fold at intervals of around 10–20 nm to form lamellae. Where branching or another irregularity of the polymer chain occurs, the direction of growth of the lamellae changes and begins to radiate out from a centre. With time and suitable temperatures, large, pyramidal structures known as spherulites develop with diameters up to 0.1 mm (Figure 4.3). As the formation of a spherulitic structure nears completion, the edges of adjacent structures begin to come into contact and even overlap each other, forming boundaries. Spherulites are often sufficiently large to interfere with light waves and cause opacity.

Figure 4.3
Spherulites are crystalline, pyramidal structures with diameters up to 0.1 mm.

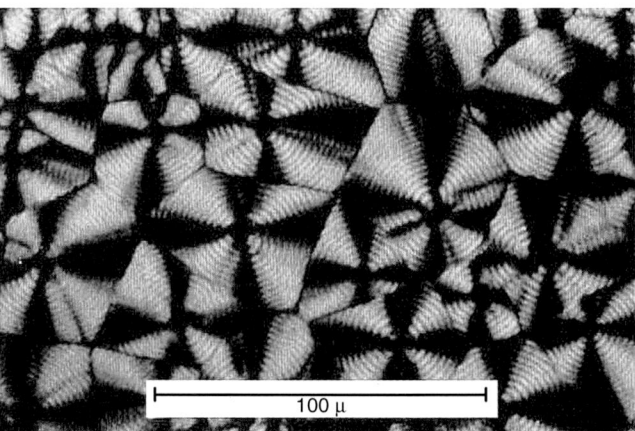

100 μ

No crystallization takes place above a polymer's melting point. If cooled slowly, polymers form many large crystals while rapid cooling results in few, small crystals. Polyethylene and poly (tetrafluoroethylene) are examples of polymers which form crystalline areas. Polycarbonate cannot form crystals, despite its regularity, due to its tetrahedral structure and very stiff polymer chain which is insufficiently flexible to change shape (Pedersen, 1999). If a polymer chain has many bulky side groups, it cannot pack sufficiently close to its neighbours to form ordered, crystalline regions, so non-ordered, amorphous regions are formed. This is demonstrated by polystyrene where the bulky benzene ring in the side chain prevents close packing.

Crystallization can also be induced by stretching rubbery polymers. In 1925, Katz identified the presence of crystals in stretched rubber bands using X-ray spectroscopy (Carraher, 2000). Rubber becomes white and opaque when stretched due to the formation of crystallites. On stretching, polymer chain segments are aligned and crystallization is induced by orientation. The crystals disappear when the stretch is relaxed. This phenomenon is used during processing of some polymers to produce films with both high strength and high clarity. Stretched poly (ethylene terephthalate) and polypropylene films contain crystals which contribute strength while allowing access to light waves.

4.1.2 Solubility of plastics

An effective solvent for plastics objects may be required as a component of an adhesion or repair treatment where the edges to be joined require etching or softening. By contrast, a poorly effective solvent for plastic may be sought as a component of a cleaning treatment where the soil is dissolved while the original plastic surface remains undamaged. In addition, selection of an effective and appropriate solvent may also depend on its evaporation rate, environmental or health and safety considerations.

A chemical is defined as a solvent for another material if the molecules of the two mix uniformly without separating. Rules of thumb such as 'like dissolves like' and an understanding of solvent 'strength' help to predict which general types of solvent are most likely to dissolve particular polymers. For solution to take place, solvent molecules must overcome the weak intermolecular attractions holding the polymer molecules together. In addition, the intermolecular interactions between solvent molecules must be separated from each other by the polymer. These two activities are only likely to take place if the intermolecular attractions of the polymer and those of the solvent are similar. A large imbalance between the two energies will favour either the polymer molecules staying together or the solvent molecules remaining close to each other, causing immiscibility.

Uneven sharing of electrons by the atoms in a molecule results in a minute imbalance between negative and positive charges, which causes the molecule to become a miniature magnet or dipole. Polymers are soluble not only when the magnitude of intermolecular forces is similar to that of potential solvents, but also when their intermolecular forces contain the same polar contributions.

Such contributions include dispersion forces, polar forces and hydrogen bonding forces. Hydrogen bonding develops in molecules where a hydrogen atom is attached to an electron hungry (electronegative) atom such as oxygen, nitrogen or halogens. The hydrogen's sole electron is drawn towards the electronegative atom, leaving the exposed positive hydrogen nucleus to attract electrons from other molecules.

The many theories behind the various models developed to calculate the solubility of polymers, and to predict the ability of liquids to dissolve them, are described clearly and in high detail by Burke (Burke, 1984). All define a term known as solubility parameter for liquids and polymers using one or more of the intermolecular force components and represent the parameter in two or three dimensions. Calculating solubility parameters is a mathematically complex process which will not be discussed here. The most widely used method today for predicting whether a polymer is soluble in a liquid was developed by Charles M. Hansen in 1966. Hansen parameters (δ) for solvents and polymers are calculated from the dispersion force component (δd), polar component (δp) and hydrogen bonding component (δh) for each using the formula:

$$\delta^2 = \delta d^2 + \delta p^2 + \delta h^2$$

A three-dimensional model is used to plot polymer solubilities by giving the coordinates of the centre of a solubility sphere based on dispersion force components, hydrogen bonding and polar components, and by plotting a radius of interaction of around 2 SI units. A sphere of solution is plotted from the coordinates and radius. Liquids whose parameters lie within the sphere for a particular polymer are likely to be suitable solvents for it (Hansen, 1971). While extensive data has been published for liquids, the number of Hansen solubility parameters for polymers is more limited (Barton, 1983). From the selected solubility parameters for liquids and polymers in Table 4.1, it is clear that the high value for water excludes it as a solvent for polymers and that polystyrene and poly (methyl methacrylate) should be soluble in acetone.

4.1.3   Chemical reactivity

The weakest chemical groups or bonds in polymer structures provide starting points for chemical reactions which result in degradation of the material. Because polymers comprise repeat chemical structures, the same groups occur at regular intervals along the chain. The factors causing degradation in various polymers and the chemical groups most affected are detailed in Chapter 6, but some general trends are presented here.

Hydrocarbon polymers such as polyethylene and polypropylene which contain only C—C and C—H bonds are largely unreactive. Similarly, polymers

Table 4.1	Hansen solubility parameters for some liquids and polymers			
	Dispersion force component d (MPa$^{1/2}$) δ	Polar component δp (MPa$^{1/2}$)	Hydrogen bonding component δh (MPa$^{1/2}$)	Hansen solubility parameter (MPa$^{1/2}$) δ
hexane	16.7	0	0	16.7
tetrahydrofuran	16.8	5.7	8.0	19.5
acetone	15.4	10.4	7.0	20.0
methanol	15.2	12.3	22.3	29.2
water	12.3	31.3	34.2	48.1
polyethylene	16.6	0	0	16.6
polystyrene	18.3	1.0	3.3	18.6
poly (methyl methacrylate)	15.7	8.2	6.8	19.0

which contain only C—C and C—F bonds such as poly (tetrafluoroethylene) are very stable and unreactive. If other carbon-halogen and carbon-hydrogen bonds are present in the same polymer, the material becomes reactive.

Polymers which contain carbon-carbon double bonds react with oxygen, ozone (O_3), hydrogen halides and halogens (such materials include synthetic rubbers). Ozone is highly reactive and causes scission at double bonds in polymer backbones, which dramatically reduces the molecular weight of the polymer. Reaction at one double bond per chain causes the average molecular weight to halve. This causes an increase in solubility and dramatic loss of physical and mechanical properties. Hydroxyl groups (OH) are highly reactive. They may be found in semi-synthetic polymers such as cellulose nitrate and acetate. Ester, amide and carbonate groups are susceptible to hydrolysis, that is addition of water molecules, which causes a reduction in molecular weight.

4.1.4   Flammability

The combustion of plastics is a complex process and only a brief overview is given here. A detailed description with data relevant to the combustion processes of common plastics may be found in specialist literature (Beyler and Hirschler, 2002). The plastics in collections which are infamous for their behaviour on combustion are cellulose nitrate and polystyrene. Cellulose nitrate is highly flammable, particularly when new and in the early stages of degradation, and burns explosively producing sparks (Figure 4.4). Polystyrene

Figure 4.4

Cellulose nitrate burns explosively and produces sparks.

produces a choking, black smoke on burning and develops an acrid odour (Figure 4.5).

A source of heat increases the temperature of the plastic which generates free radicals. Decomposition of the polymeric component accelerates with temperature resulting in the production of volatile gases and liquids, charred solids and smoke. Some of the products will slow the rate of decomposition while others accelerate the process. Ignition takes place when both volatile gases and oxygen are present in sufficient quantities above the ignition temperature. The quantities of oxygen required for ignition vary between polymers. Polyethylene, polypropylene and polystyrene require 17–20 per cent whereas nylons, phenol-formaldehyde and poly (vinyl chloride) require 25–40 per cent

Figure 4.5
Polystyrene burns with
a sooty flame and a
choking black smoke.

for ignition and continued burning (Brydson, 1999). Combustion follows ignition. If the burning gases feed back sufficient heat to the plastic to generate
flammable volatiles, the process is self-sustaining and burning continues.

More fatalities occur via suffocation by smoke or toxic gases emitted during a fire than by burning. It is therefore worrying to consider that some fire
retardants incorporated into polymers act by increasing the amount of smoke
produced while slowing the rate of flame development.

4.2   Optical properties

The great economic and cultural success achieved by plastics can be attributed
largely to their optical properties. The ability of semi-synthetics to mimic the
rich colours and opacity or transparency of natural materials endeared them

to the public around 1900. The glass-like clarity of acrylics resulted in many important military applications in World War II, including windscreens for vehicles and aircraft and as colourful advertising and information signs in the post-war period. Polyester fizzy drink containers are visually attractive because the contents can be seen.

For polymers without additives and impurities, the degree of translucency is influenced primarily by the extent of crystallinity. In highly crystalline polymers such as high density polyethylene and polypropylene, the polymers fold to form orderly crystalline and the polymers appear cloudy or opaque. Crystalline spherulites are often approximately the same size as the wavelength of visible light, causing the light to scatter, while polymers containing numerous, small crystallites are more transparent. Amorphous polymers can transmit 92 per cent of incident light while crystalline materials are only transparent as very thin films.

As crystalline materials melt, their appearance transforms from opaque to transparent because the ordered structure is lost. Highly amorphous polymers, including acrylics, polycarbonate, and polystyrene do not form crystals, so are transparent (Figure 4.6). An exception is crystalline polyester poly (ethylene terephthalate) used in fizzy drinks bottles, which is transparent because its crystals are too small to interfere with light waves. Fillers and additives usually decrease the light transmission of a plastic by scattering incident light.

Figure 4.6
The extent of crystallinity and size of crystals where present affect the transparency of polymers. Polyethylene (far left) is highly crystalline and cloudy; polystyrene (centre) is amorphous and transparent; and poly (ethylene terephthalate)'s crystals are too small to interfere with light (far right).

When light is transmitted from air into the interior of transparent plastics, it slows and, as a result, is bent at the air/plastic interface, a phenomenon known as refraction. The degree of bending is dependant on the wavelength of light, so that light of different colours will be deflected by different amounts as they emerge after passing through plastics. The refractive index (or index of refraction) of a material (n) is defined as the ratio of the velocity of light in a vacuum (c) to its velocity in a material (v).

$$n = \frac{c}{v}$$

If the material has a refractive index similar to that of air (1.00), it appears transparent. Silica glass (1.46), polycarbonate (1.59), polystyrene (1.60), and acrylics (1.49) all have refractive indices similar to that of air. When making transparent parts in plastic, designers must ensure that materials with the same index of refraction are used in distinct areas, rather than distributed throughout the design in order to avoid light bending in different directions and undergoing distortion (Table 4.2).

Table 4.2 Refractive indices of polymers without additives or fillers	
Polymer type (without additives or fillers)	Refractive index at 20°C (n)
poly (tetrafluoroethylene)	1.3500
cellulose acetate	1.4757
polyethylene	1.5100
cellulose nitrate	1.5100
plasticized poly (vinyl chloride)	1.5390
epoxy	1.5500
polyamide	1.5650
polycarbonate	1.5860
polystyrene	1.5894
phenol formaldehyde	1.7000

4.3   Physical properties

Physical characteristics of polymers are dependant on their molecular weights, molecular shapes and structures. The physical properties of greatest interest to conservation professionals are those used to determine plastics' reaction to outside stimuli and environment, their mechanical behaviour and their response to heat. These properties described here under the categories density, electrical, mechanical and thermal.

4.3.1   Density

The density of polymers is related to the mass of their molecules and the way they pack. Hydrocarbons (those composed only of hydrogen and carbon), such as polyethylene and polypropylene, do not contain heavy atoms compared to those containing halogens such as PVC and poly (tetrafluoroethylene). The polymer chains in amorphous materials do not pack closely. Amorphous hydrocarbon polymers have low densities ($0.86–1.05\,\mathrm{g\,cm^{-3}}$) compared with a density for highly amorphous PVC of $1.4\,\mathrm{g\,cm^{-3}}$.

Polymers containing crystalline structures pack more closely and attain higher densities than amorphous materials. Crystallinity contributes to the high densities of crystalline poly (tetrafluoroethylene) (ca. $2.2\,\mathrm{g\,cm^{-3}}$) and poly (vinylidene chloride) (ca $1.7\,\mathrm{g\,cm^{-3}}$). The conformation of molecules in the crystalline structures also affects the density. Polyethylene adopts a planar conformation while polypropylene molecules have a helical conformation in the crystalline regions. Helices require more space than planar forms, resulting in a lower density for polypropylene than polyethylene (Brydson, 1999).

4.3.2   Electrical properties

In general, polymers are poor conductors of electrical charge. Resistivity values for polymers are of the order 10^{12} Ohm cm compared with 10^2–10^4 Ohm cm for semiconductors and 10^{-5} Ohm cm for metals. Because resistivity is the inverse of conductivity, these values suggest that pure polymers can be considered electrical insulators. Incorporation of selected fillers, graphite in particular, increases the conductivity of plastics. In addition, transportation of charge in plastics can occur via their surfaces. This is commonly known as static. As an example, polyesters contain ester linkages comprising oxygen atoms with mobile electrons. At surfaces, these polar groups can attract moisture or statically charged surface contaminants, including dust, which disfigure plastics' surfaces and have the potential to accelerate hydrolytic breakdown (Walton and Lorimer, 2003).

4.3.3   Permeability

Plastic films are often used as barriers to shield the enclosed materials from contact with oxygen or water vapour (Figure 4.7). However, not all polymers are equally permeable to gases. Permeation of a gas or liquid through a film of plastic is a three-stage process. It involves a solution of a small molecule in the polymer, migration or diffusion through the polymer at a rate controlled by the concentration gradient, and emergence of the molecule at the other side. Permeability is the product of solubility and diffusion (Brydson, 1999).

Diffusion rate is affected by the size of both the diffusing molecules and the gaps between the polymer molecules. The size of the gaps depends largely on the physical state of the polymer under conditions of use, whether glassy, rubbery or crystalline. The physical state affects the extent of packing of polymer

Figure 4.7
The low permeability of poly (ethylene terephthalate) to carbon dioxide and oxygen has been applied to beer bottles. Carbon dioxide in the beer escapes only slowly and oxygen, which would spoil the flavour of beer, cannot permeate in from the surroundings.

chains, their mobility and the extent of movement between them. Crystalline structures have a much higher degree of molecular packing than amorphous areas, so can be considered as almost impermeable to gases and liquids. As a result, crystalline polymers tend to resist diffusion more than either glassy polymers or rubbery ones (Table 4.3). Permeability of polymers is measured by placing a film of the test material between two chambers. Oxygen, air or vapour at high pressure is applied to one side of the test film and the time taken for the pressure to equalize in both chambers is determined.

4.3.4   Mechanical properties

The mechanical properties of polymers are highly dependent on their molecular weight. Plastics based on polymers with long polymer chains are mechanically

Table 4.3 Permeability to oxygen and water vapour for various polymers			
Polymer	Permeability to O_2 at 30°C water vapour (90% relative humidity) at 25°C		State of polymer
	$(P \times 10^{10} \, cm^3 \, s^{-1} \, mm \, cm^{-2} \, cmHg^{-1})$		
poly (ethylene terephthalate)	0.22	1300	crystalline
nylon 6	0.38	7000	crystalline
PVC (unplasticized)	1.20	1560	crystalline
cellulose acetate	7.80	75000	glassy
polyethylene	55	800	crystalline
polystyrene	11	1200	glassy
polybutadiene	191	no data available	rubbery

stronger than those containing shorter chains. Plastics can be formulated to produce a wide range of mechanical properties. They can withstand a similar maximum stress to pine wood or silica glass before failing or may extend up to 1000
per cent at room temperature which is 10 times that possible for metals (Table
4.4). By contrast, the mechanical and dimensional properties of plastics are
more sensitive to change in temperature than other materials. Determination of

Table 4.4 Tensile strength at break (ultimate strength) for various materials at 20°C	
Material	Tensile strength at break (MPa)
carbon fibre	5650
silicon carbide	3440
stainless steel – cold rolled	860
bone (leg)	130
nylon 6,6	76–94
pine wood (parallel to grain)	40
glass (silica)	27–62
polypropylene	20–41
high density polyethylene	20–30
marble	15
concrete	3

the stress–strain characteristics of poly (methyl methacrylate) shows it to be brittle and inflexible at 4°C while exhibiting plastic deformation and flexibility at 50°C (Callister, 1994).

On tensile testing plastics exhibit distinct tendencies which can be grouped (Figure 4.8). Rigid plastics such as polystyrene, poly (methyl methacrylate) and polycarbonate withstand considerable load or stress but little elongation before breaking. They are strong but not tough, where toughness is a measure of the energy a material can absorb before it fails. Toughness is quantified by determining the area under the stress–strain curve. Synthetic fibres such as polyamides, nylons and very high molecular weight polyethylene are poorly flexible but can withstand very high loads prior to sudden breaking.

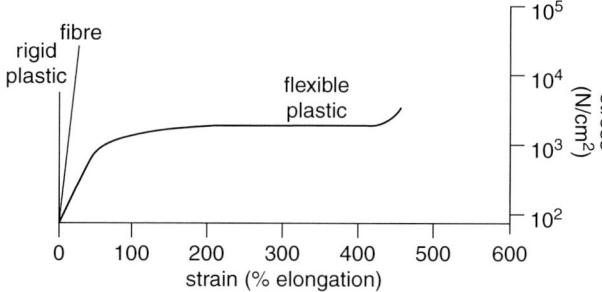

Figure 4.8
Typical stress–strain curves for plastics.

Flexible plastics including polyethylene and polypropylene cannot withstand as much load as rigid plastics but have superior abilities to stretch and extend. Polyethylene carrier bags can be stretched and will recover their original length initially. On stretching further, they extend and form a neck, becoming thinner until failure (Figure 4.9). Elastomers including polyisoprene and polybutadiene, cannot withstand high loads (low moduli) but have very high extensions and, on release of load, resume their original lengths.

Combining one or more polymers chemically by copolymerizing or physically blending them results in a new material with some properties from each of the starting ingredients. A polymer's mechanical properties can also be modified with additives, particulate fillers and reinforcing fibres comprising glass, metal or carbon. Properties can vary with direction of measurement within the plastic, particularly if fibres are present. Fillers such as talc or ground calcium carbonate increase stiffness while plasticizers decrease tensile strength and increase flexibility.

On a molecular level, strain–stress measurements for polymers vary with crystallinity, chain entanglement and other structural factors. Crystalline polymers contain areas where molecular chains are closely packed. Weak secondary intermolecular bonds connect adjacent chain segments and inhibit inter-segmental motion. Amorphous areas contain a lower density of secondary

Figure 4.9

Polyethylene stretches and extends to form a neck (centre) prior to breaking (right).

bonds because of the reduced order and reduced proximity of polymer chains. As a result, polymers with higher percentages of crystallinity generally exhibit superior mechanical properties compared with amorphous materials of a similar molecular weight. Amorphous polypropylene is more flexible than crystalline polypropylene because polymer chains in the former interact less with each other than in the latter. Linear polymers tend to show higher tensile strengths than branched or networked materials while the presence of branching promotes toughness in polymers.

Rotational flexibility is also a function both of monomer structure and the type of chemical groups present in the polymer chain (Carraher, 2000). The regions of polymer chains containing stiffening groups are rigid. Such groups include double bonds (C=C), phenylenes (benzene ring structure) and carbonyls (C=O). Where stiffening groups are present, flexibility may be increased if softening methylene groups (CH_2) or oxygen atoms are inserted between the stiffening components in a chain. The flexibility of the polymer increases with the number of softening groups. Poly (ethylene terephthalate) (PET) is stiffer and has a higher melting point than poly (butyl terephthalate) (PBT) because the former has fewer methylene groups between the stiffening phenylene and carbonyl groups. The same is true where a bulky side group is present, for example the benzene ring present as a side group in polystyrene renders the polymer inflexible compared with polyethylene, which does not have large side groups.

4.4   Thermal properties

One method of classifying plastics is by their response to heat. Thermoplasts, also known as thermoplastic polymers, soften and liquefy on heating and harden again when cooled. The process is reversible and can be repeated. On heating, the weak secondary bonds between polymer chains are broken, which facilitates relative movement between the chains. If the molten polymer is further heated until the primary covalent bonds also break, degradation of the thermoplast follows. Thermoplastic polymers are linear or exhibit branching with flexible chains and include polyethylene, polystyrene and polypropylene (Figure 4.10).

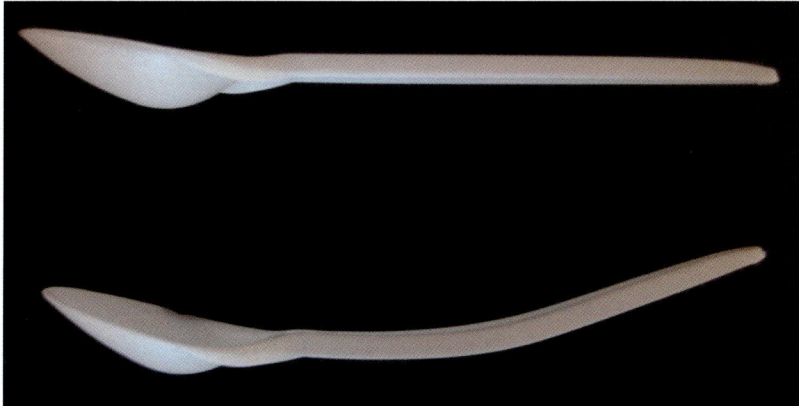

Figure 4.10
Thermoplastic polystyrene spoon before (top) and after (bottom) stirring a hot cup of coffee. The deformation caused by heating can be reversed by repeating the heating process.

Thermosets comprise crosslinked and network polymers such as phenolics and polyesters. Thermosets, also known as thermosetting polymers, become permanently hard on heating and do not soften again. At a molecular level, crosslinks form between 10–50 per cent of the adjacent polymer chains so that they can resist the vibrational and rotational chain movements induced on heating. Thermoset polymers are harder, stronger and more brittle than thermoplastics.

4.4.1   Thermal expansion

Plastics, like all materials, experience a change in length, area or volume with temperature. On an atomic level, thermal expansion is caused by an increase in the average distance between atoms. Heating increases the vibrational energy and vibrational amplitude of atoms, so that the maximum interatomic distance can also increase. Both expansion on increasing temperature and shrinkage on cooling are fully reversible. The coefficient of linear expansion for plastics varies between types but ranges from 50×10^{-6} to $300 \times 10^{-6}(°C)^{-1}$, which is at least 10 times higher than that for metals (5×10^{-6} to $25 \times 10^{-6}(°C)^{-1}$) or ceramics ($0.5 \times 10^{-6}$ and $15 \times 10^{-6}(°C)^{-1}$). Linear and branched polymers exhibit the highest coefficients of expansion because the secondary bonds prevalent in them are weak. The coefficient of expansion decreases with increased crosslinking and is therefore lowest in thermosetting network polymers such as phenol-formaldehydes (Bakelite).

The variation in coefficients of thermal expansion between polymers gives difficulties where materials are composites or where several different material types are in close contact. On cooling, the differential shrinkages may generate stress between the materials, causing them to fail or to come apart from each other.

4.4.2 Phase transitions on heating

Melting is the transition from solid to liquid which occurs in crystalline polymers on application of sufficient heat. Melting corresponds to the transformation of a solid material with an ordered structure of aligned molecular chains to a viscous liquid in which the structure is highly random. At ambient temperatures, atoms vibrate with small amplitudes relatively independently of one another. Weak secondary bonds form between adjacent polymer chains. On heating, atoms vibrate with greater amplitudes and become coordinated so that movement of entire polymer chains is induced. At melting temperature (Tm) movement of polymer chains is sufficient to destroy a large proportion of the secondary bonds, resulting in a disordered structure and a mobile liquid.

Melting is known as a first order transition. Unlike crystalline polymers, amorphous polymers do not melt when heated but undergo a dramatic reduction in flexibility when they are cooled below a characteristic temperature known as the glass transition temperature (Tg) (Figure 4.11). Below Tg, the

Figure 4.11

Heating crystalline polymers (1) causes melting at a characteristic temperature, Tf (also known as Tm), whereas amorphous polymers (2) undergo a dramatic physical change in mechanical properties from rubbery to glassy physical at a characteristic glass transition temperature, Tg.

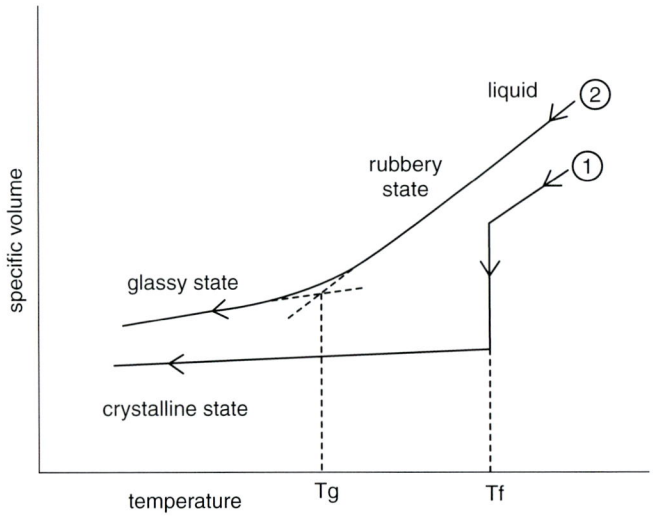

polymer chain segments are immobilized, imparting rigidity and glass-like properties. Above Tg, polymers are rubbery and flexible.

The best known illustration of this phenomenon is the cooling of rubber tubing in liquid nitrogen (−196°C). The Tg of natural rubber is −70°C. Above −70°C, the material is flexible and readily absorbs shock. When cooled below its Tg, rubber becomes stiff and shatters on impact because the polymer chains no longer have sufficient energy to absorb energy by rotating, so the impact breaks bonds in the polymer structure. In addition to stiffness, volume occupied by the material, dielectric properties, gas permeability, and refractive index of amorphous polymers change at Tg.

The phenomenon of glass transition is considered a second order rather than a first order transition because the polymer remains a solid both above and below the Tg, and does not change phase. The temperature at which the change between glassy and rubbery behaviours occurs is not always distinct but may be the middle temperature in a glass transition region. Values for Tg are usually 33–100 per cent lower than melting temperature. Symmetrical polymers such as high density polyethylene exhibit the greatest differences between melting point and Tg. No polymer exhibits 100 per cent crystallinity but contains crystalline and amorphous regions and has both Tm and Tg values (Table 4.5). The crystalline areas melt on heating while the amorphous regions undergo glass transition.

Table 4.5 Approximate glass transition and melting temperatures for polymers		
Polymer	**Glass transition temperature Tg (°C)**	**Melting temperature Tm (°C)**
polyethylene	−120	137
polypropylene	−15	176
nylon 6	50	215
poly (vinyl chloride) – unplasticized	80	90
polystyrene	100	240
poly (methyl methacrylate)	100	160
cellulose triacetate	107	300

The values of both Tg and Tm are affected by polymer structure, in particular by chain flexibility, steric hindrance, side group effects, symmetry, polarity and the inclusion of plasticizers in the formulation. Rigid polymer chains give fewer opportunities for bond rotation than flexible backbones and the result is an increase in Tg. An example is polyethylene which has a very flexible backbone and a Tg of $-120°C$ compared with polycarbonate which contains two phenylene groups (benzene ring structure) and one carbonyl ($C=O$) per repeat unit and has a Tg of $+150°C$ because less 'cooling' is required to immobilize its backbone (Figure 4.12). As with Tg, the stiffer the polymer backbone, the higher Tm. More energy is required to separate molecules which are stiff than those which are flexible.

Figure 4.12

Influence of flexibility on glass transition temperature.

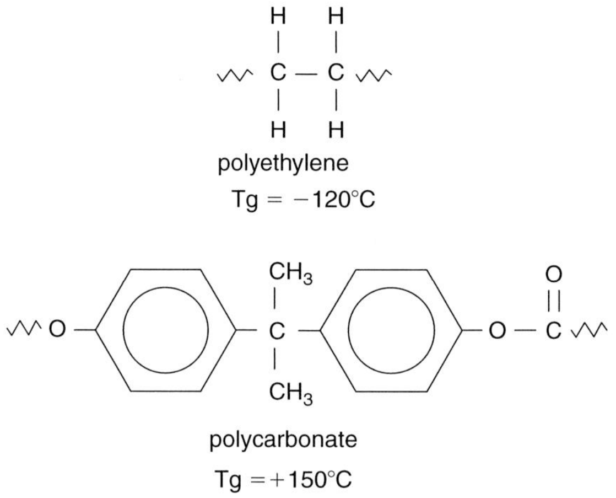

polyethylene

Tg = $-120°C$

polycarbonate

Tg = $+150°C$

The presence of large side groups attached to the backbone inhibits rotation and results in higher Tg and Tm values than those for polymers with small side groups. Increasing Tg values for the series polyethylene ($-120°C$), polypropylene ($-15°C$) and polystyrene ($100°C$) may be attributed to their side groups which are hydrogen, methyl and benzyl respectively. Less 'cooling' is required to immobilize polystyrene than polyethylene. In addition to the size of the side group, its flexibility also contributes to Tg. A similar pattern is seen with melting points of polyethylene ($+137°C$), polypropylene ($+176°C$) and polystyrene ($+240°C$) which increase with size of side groups.

Larger side groups may be more flexible than small ones, requiring more energy to stop them rotating so that they can adopt a glassy form. This compensates for their size. An example is the series of acrylates. Despite increasingly large side groups, the Tg of methacrylates decreases down the series because the flexibility of the side group also increases on going from poly (methyl methacrylate), which has a Tg 100–120°C, to poly (ethyl methacrylate) with Tg 65°C to poly (butyl methacrylate) with Tg 20°C (Figure 4.13).

Increase in polarity causes an increase in both Tg and Tm. Non-polar polypropylene has a Tg of $-14°C$, moderately polar PVC has a Tg $+87°C$ and highly polar polyacrylonitrile has Tg $+103°C$. The relationship between polarity and Tg may be attributed to the increase in polar bonding between polymer chains as the methyl group in polypropylene is replaced by the chlorine

$+CH_2-C+_n$ CH_3 ... C=O ... O ... CH_3
poly (methyl methacrylate)
Tg = 100–120°C

$+CH_2-C+_n$ CH_3 ... C=O ... O ... CH_2 ... CH_3
poly (ethyl methacrylate)
Tg = 65°C

$+CH_2-C+_n$ CH_3 ... C=O ... O ... CH_2 ... CH_2 ... CH_3
poly (propyl methacrylate)
Tg = 35°C

$+CH_2-C+_n$ CH_3 ... C=O ... O ... CH_2 ... CH_2 ... CH_2 ... CH_3
poly (butyl methacrylate)
Tg = 20°C

Figure 4.13
Tg of methacrylates decreases down the series, despite increasingly large side groups, due to their increasing flexibility.

group in PVC and the cyano (CN) group in polyacrylonitrile. Increased bonding restricts rotation around the backbone, so that less energy is required to 'freeze' motion in the most polar polymer. The high cohesive energies and intermolecular forces found in polar polymers also result in increased melting points compared with non-polar materials.

Adding between 30–40 per cent by weight phthalate plasticizer to PVC reduces its Tg from around 80°C to −40°C. Plasticizers are low molecular weight molecules compared with polymers. They are often oily liquids which increase mobility by separating the polymer chains. Plasticizers improve the ability of polymer chains to move past each other. This increase in mobility requires the polymer to be cooled to lower temperatures than without a plasticizer, to immobilize the chains.

Tables of optical, physical and thermal and chemical properties of the plastics most frequently collected may be found in Appendix 1.

References

Barton, A. F. M. (complier) (1983). *Handbook of solubility parameters*. CRC Press.

Brydson, J. A. (1999). *Plastics Materials*, 6th edition. Butterworth-Heinemann.

Beyler, C. L. and Hirschler, M. M. (2002). Thermal decomposition of polymers. In *SFPE Handbook of Fire Protection Engineering*, 3rd edition. (P. J. DiNemmo, ed.) pp. 1/110–1/131, NFPA, Quincy, MA, USA.

Burke, J. (1984). Solubility parameters: theory and application. *The Book and Paper Group Annual*. The American Institute for Conservation (3) [online]. Available from: http://aic.stanford.edu/sg/bpg/annual/v03/bp03-04.html [Accessed 20 December 2006].

Hansen, C. M. (1971). Solubility in the coatings industry. *Sksnd. Tidskr. Faerg. Lack.*, 17, 69.

Callister, W. D. Jr. (1994). *Materials science and engineering – an introduction*, 3rd edition. Wiley, New York.

Carraher, C. E. Jr. (2000). *Polymer Chemistry*, 5th edition. Marcel Dekker Inc., New York.

Pedersen, L. B. (1999). *Plast og Miljø (Plastics and Environment, in Danish)*. Teknisk ForlagA/S, Copenhagen.

Walton, D. and Lorimer, P. (2003). *Polymers*. Oxford University Press, Bath.

Identification of plastics in collections

Summary

Chapter 5 discusses the reasons and techniques for identifying plastics in collections. In addition to establishing the age and technological history of a plastic, identification indicates the factors and pathways by which degradation is likely to occur and thus helps the conservator to develop a treatment strategy. Simple tests enable identification of the polymer type while instrumental techniques are necessary if the various components are to be characterized. This chapter describes simple and non-destructive tests, simple and destructive tests, chemical spot tests and instrumental analytical techniques which are applied to plastics in collections today.

Identifying the types of plastics present in objects gives information about how, when and why they were produced. In addition, for a conservator, knowledge of the materials present provides information about the most likely degradation pathways. This knowledge is necessary to develop appropriate conservation strategies. If degradation results in the production of acidic gases that accelerate the rate of degradation of the plastic, as in the case of cellulose nitrate, enclosing it in a sealed container would not be an appropriate or effective conservation strategy. However, such an approach would be suitable for a plastic which deteriorates due to loss of volatile additives such as poly (vinyl chloride). Selection of solvents for cleaning, adhesives for repair and consolidants for stabilizing plastics surfaces cannot be carried out without risking permanent damage to objects unless the substrate's identity is established.

For a collector or curator, it is important to know if the jewellery or other piece is made of plastic rather than a natural polymer and if the plastic is in its original form. The term 'Fakelite' seems to have been used first in 1988 to describe Bakelite buttons that had been glued together to form a brooch, which was being falsely described as Art Deco jewellery from the mid 1920s to increase its commercial value (Davidov and Dawes, 1988). This material is considered as reworked Bakelite. Today the term 'Fakelite' is used to describe jewellery newly manufactured from either urea-formaldehyde, melamine-formaldehyde or other plastics, which is being falsely represented as vintage Bakelite (phenol-formaldehyde) and comes from several sources including Taiwan. Misrepresentation has contributed to a fall in the price of Bakelite jewellery in the USA.

The conservation profession use the same techniques for identification as the plastics industry, although the purpose of examination may differ. In addition to confirming that the polymer type is that expected from the description, function and age of a plastic, industry is also concerned that the physical and chemical properties of plastics match the required performance or quality. Industrial analysis is usually concerned with quantifying components of plastics such as additives, while conservation professionals are more concerned with qualitative examination.

Because test methods may vary substantially between laboratories, organizations such as the American Society for Testing and Materials (ASTM) and the British Standards Institute (BSI) have developed standardized tests that are used throughout the plastics industry. Methods of identification may be divided into simple tests, including examination of appearance, measurement of physical and chemical properties, and those involving more complex, instrumental techniques. Many conservators and private collectors do not have access to analytical instruments, so employ simpler identification techniques. Selection of an examination technique is also dependent on whether the plastic can be sampled without damaging or changing its significance. Sampling of cultural materials is not always permitted.

It is easier to characterize unknown material if its historical and technological backgrounds are known. The development of semi-synthetic polymers in the second half of the nineteenth century can be followed using patents (Fernandez-Villa and Moya, 2005). Development of cellulose nitrate was prompted by the need to find substitutes for natural materials including tortoiseshell, ivory and ebony, which were very expensive. By 1858, approximately 8 per cent of British patents concerned the synthesis or moulding of semi-synthetics.

For more recent synthetic polymers, interviews with the artist can provide information about materials and technique used, although they do not always replace analysis. The date of manufacture or, if unavailable, the date of collection, can provide a starting point in the identification of plastics. If a plastic was manufactured before around 1905, it is likely to be a semi-synthetic or natural material, rather than a fully synthetic plastic. The period between 1939 and 1960 saw a dramatic increase in production of polystyrene, poly (vinyl chloride), nylon, acrylics and polyethylene and the phasing out of semi-synthetics. If a plastic was manufactured before the 1940s, it could not have been shaped by injection moulding, recognized by a small imperfection due to the filling hole, but by another process. Styles are not always reliable sources for identifying age or material because the same mould can be used for many years and popular styles are often revived.

Plastics comprise one or a mixture of polymers, additives and sometimes reinforcing materials. Because it occupies most of the formulation and therefore controls most of the chemical properties of the plastic, the polymer is the component most frequently used to identify the plastic. Another reason for basing identification on the polymer component as opposed to the additives is to reduce the number of variables. While there are approximately 50 different groups of polymer, there are thousands of different additives and the same additive can be included in many different plastic formulations. As an example, di-n-butyl phthalate is used as a plasticizer for many plastics including PVC and epoxies, so its identification gives little information about the specific plastic present.

Identification of polymers is a complex process. Additives change the physical properties of polymers and their presence may confuse the results of tests to examine flexibility, hardness and density. Degradation of polymers changes their physical and chemical properties compared with new materials. Degraded polymers are often stiffer and darker than new materials, which is likely to confuse identification based on appearance. Their surfaces may become tacky due to migration of additives or disrupted by the formation of cracks.

Before embarking on tests involving the use of solvents or other chemicals, some of which are described in this chapter, it is important to consider the health and safety risks associated with such materials and how to minimize them. In the UK, the use of hazardous materials is regulated by the Control of Substances Hazardous to Health Regulations 2002 (CoSHH). The main objective of the Regulations is to reduce occupational ill health by setting out a simple framework for controlling hazardous substances in the workplace (Health and Safety Executive, 2007).

A fundamental requirement of the CoSHH regulations is that the exposure of employees to hazardous substances should be prevented, or, where this is not reasonably practicable, adequately controlled. Exposure to harmful materials can occur by inhalation, by ingestion or by absorption through the skin but inhalation is usually the main route of entry into the body. The Health and Safety Commission sets 'Occupational Exposure Limits', or concentrations of substances in the air at or below which exposure control is considered to be adequate. The values for exposure limits are listed on the Oxford Physical and Theoretical Chemistry Laboratory home page (Oxford Physical and Theoretical Chemistry Laboratory, 2006). Other countries have their own regulations for minimizing risks to users of chemicals and it is recommended that these are consulted before carrying out the tests described here.

This chapter describes simple and non-destructive tests, simple and destructive tests, chemical spot tests and instrumental analytical techniques which are applied to plastics in collections today.

5.1 Simple non-destructive identification techniques

Appearance and odour are the major non-destructive simple tests used to identify plastics.

5.1.1 Appearance

Colour and opacity can only be a guide to the type of plastic present because fillers and colouring materials can change the appearance and surface texture of a polymer considerably. Plastics produced at the start of the twentieth century were often formulated and shaped to resemble expensive and rare natural materials resulting in deceptive appearances (Figure 5.1). There are many examples of plastic jewellery and decorative pieces which are registered as containing amber or copal beads or ivory components in museum collections, based on appearance alone (Green and Bradley, 1988) (Coxon, 1993). The material is usually only investigated further when unexplained corrosion of metal components in the vicinity takes place.

Figure 5.1

Cellulose nitrate was often formulated and shaped to imitate tortoiseshell as shown by these two hair combs. The comb on the right is real tortoiseshell while that on the left is cellulose nitrate and dates from 1910.

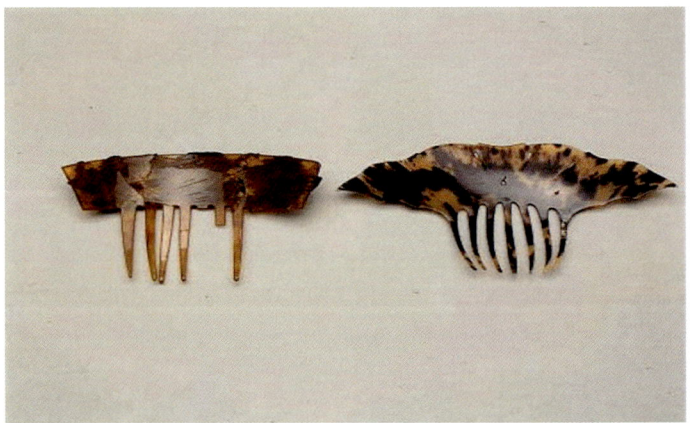

In general, Bakelite (phenol-formaldehyde) is black, dark brown, green or red and mottled, although if used unfilled as a casting resin instead of being shaped by moulding, it can be found in paler colours and is used to imitate jade, amber and onyx (Coxon, 1993). Some polymers have restricted colour ranges, particularly thermosets. Polyethylene and polypropylene always appear slightly cloudy in film or sheet form due to their high crystallinity, which diffuses light, while most other colourless plastics, particularly polyesters and polycarbonate, are crystal clear. The presence of reinforcing materials increases the stiffness and roughens the surface profile of plastics. Incorporated wood flour and mascerated paper impart a mottled appearance to phenol-formaldehydes. Glass and carbon fibres used to increase the tensile strength of polyesters and epoxies both stiffen the polymers and impart an uneven texture.

Markings on plastics may provide information about their country of manufacture, the production company (via a trademarks), design and whether the type of plastic present can be recycled or is suitable for use in contact with food (Figures 5.2 and 5.3). If a date is present in association with a company's name, it refers to the establishment of the company and not the date of production of the object. Between 1860 and 1915, British patent numbers restarted every January with number 1 followed by the year. In 1916, number 100 001 was issued and the sequence continued ignoring year ends. In 1920, number 136 852 was issued, in 1930 number 323 171, in 1940 number 516 338, in 1950 number 634 001, in 1960 number 826 321, in 1970 number 1 175 851 and in 1980 number 2 050 131 (Williamson, 1999).

Figure 5.2
Trade mark, design number and country of manufacture on the base of a nailbrush from the 1950s.

Figure 5.3
Trademark and country of manufacture on a fishing box from the 1990s.

To assist recycling of plastics, the American Society of the Plastics Industry (SPI) developed a coding system in the mid 1990s by which packaging plastics could readily be identified. The codes or universal recycling symbols were also adopted by the Plastics Manufacturers in Europe (APME). Codes comprise a triangle with a number from 1 to 7 at the centre, or 01 to 07 for the German market, each of which refers to a particular plastic (Table 5.1). An abbreviation for the plastic is placed under the triangle but, unlike the number, it is not an international standard. Because compliance in labelling is voluntary, not all plastics are marked with the code.

Table 5.1 Universal recycling symbols for plastics		
Universal recycling symbols	**Abbreviation for plastic used in Europe/USA**	**Description of plastic**
1	PET/PETE	*poly (ethylene terephthalate)*, e.g. fizzy drink bottles and oven-ready meal trays
2	PE-HD/HDPE	*high density polyethylene*, e.g. bottles for washing-up liquids
3	PVC/V	*poly (vinyl chloride)*, e.g. photograph album pages, flexible toys
4	PE-LD/ LDPE	*low density polyethylene*, e.g. carrier bags, black rubbish bags
5	PP/PP	*polypropylene*, e.g. margarine tubs, microwaveable meal trays, archive quality photograph pockets
6	PS/PS	*polystyrene*, e.g. yogurt pots, foam food dishes, hamburger boxes, vending machine coffee cups, plastic cutlery, protective packaging for electronic goods
7	O/OTHERS	*any other plastics* that do not fall into any of the above categories, e.g. melamine-formaldehyde, often used in plastic plates and cups or mixtures of recycled thermoplastics

5.1.2 Odour

The odour of plastics is frequently used as a simple aid to identification. Since heat increases the concentration of volatile materials, plastics are either rubbed with a clean cotton cloth (close to the surface to generate frictional heat) just prior to sniffing or tested in combination with a heating test. Collectors who prefer not to take samples or risk mechanical damage to an object often use hot water as a source of heat. The edge of an object is held under a hot, running tap for up to 30 seconds before smelling. The smell is usually generated by volatilizing a monomer, plasticizer or degradation product. An odour of vinegar is typical of cellulose acetate, that of muscle rub ointment or balm (camphor) suggests cellulose nitrate, and a smell similar to that of a new car indicates plasticized PVC.

A phenolic odour, similar to that of antiseptic soap indicates the presence of Bakelite (Parry, 1996). A fishy odour is often produced by warming melamine- or urea-formaldehydes. Unfortunately, description of odour is very subjective and it is difficult for testers to agree (Table 5.2). Although the odour of warm

Table 5.2 Odour of plastics when warmed	
Plastic type	**Odour when warmed to 50–60°C**
acrylics	sweet, fruity
casein	burnt milk, hair
cellulose acetate	vinegar, burning paper
cellulose nitrate	camphor (1,7,7-trimethyl-bicyclo [2,2,1] hepta-2-one), muscle rub
melamine-formaldehyde	fish
nylon	burnt hair or wool, celery
phenol-formaldehyde	antiseptic soap
polyester	raspberry jam, cinnamon, burning rubber
polyethylene	wax, candles, paraffin
polypropylene	wax, candles
polystyrene	styrene, acrid
polyurethane	acrid, stinging odour, faint apple
poly (vinyl chloride)-unplasticized	acrid, chlorine, aromatic
poly (vinyl chloride)-plasticized	sweet, new car seats
silicone	none
urea-formaldehyde	formaldehyde, preserving fluid

polyethylene is always described as wax, candles or paraffin, that of polyester has been described both as raspberry jam and cinnamon, smells which differ greatly in my opinion (Williamson, 1999; Texloc Closet, 1997). It has been proposed that a series of reference samples would improve reproducibility of this test (Coxon, 1993). That is possible, but the variation between testers' written or verbal descriptions would remain.

5.2   Simple destructive identification techniques

In general, simple observation and feel alone are insufficient to reveal the chemical nature of a three-dimensional plastic object. Identification schemes have been developed based on simple tests including the behaviour of polymers on impact, in liquids (flotation and dissolution) or in a flame. Such tests provide a rough, qualitative assessment of simple homopolymers but are not effective at identifying copolymers or polymer blends (Braun 1996; Cloutier and Prud'homme, 1993).

5.2.1   Density

Determining the density of plastics has been used as a simple, non- or semi-destructive technique for their identification. Density is dependant on the weight of molecules and the way they pack. Hydrocarbons contain relatively light atoms, so have low densities, between $0.85 \mathrm{gcm}^{-3}$ and $1.05 \mathrm{gcm}^{-3}$ (Brydson, 1999). Polymers which contain heavier atoms such as chlorine in poly (vinyl chloride) have a higher density up to around $1.4 \mathrm{gcm}^{-3}$. Crystalline polymers have a higher density than amorphous due to the more efficient packing available to the former. The presence of fillers and reinforcing material will increase the density compared with the pure polymer (Table 5.3).

The position of a sample of plastic in a beaker of a test liquid with known density at a particular temperature is related to the density of the plastic. The density of water at 20°C is $1 \mathrm{gcm}^{-3}$. If a small sample floats on the surface of the water, it has a density lower than $1 \mathrm{gcm}^{-3}$ at the same temperature. If it is suspended, the density of the plastic is $1 \mathrm{gcm}^{-3}$ and if it sinks the plastic has a higher density. Polyethylene, polypropylene and polystyrene float on water while other plastics sink (Figure 5.4). Samples of plastic films may be taken using a hole punch, but this technique is unsuitable for foams or blocks of material (Williams et al., 1998). Other fluids used include saturated sodium chloride ($1.20 \mathrm{gcm}^{-3}$), saturated magnesium chloride ($1.34 \mathrm{gcm}^{-3}$), saturated calcium chloride ($1.45 \mathrm{gcm}^{-3}$) and saturated zinc chloride ($1.57\text{--}2.00 \mathrm{gcm}^{-3}$).

The flotation test is a rough method to identify plastics since results are dependent on the physical form of the material. Foams contain cells filled with air, so their densities will be lower than a solid block of identical dimensions of the same type of plastic. Water-sensitive or absorbent fillers such as paper

Table 5.3 Specific gravity of plastics	
Plastic type	Specific gravity (ratio of the mass of a material to the mass of an equal volume of water at 4°C)
acrylics	1.16–1.20
casein	1.26
cellulose acetate	1.25–1.35
cellulose nitrate	1.34–1.38
melamine-formaldehyde	1.50
	1.80–2.10 (filled with paper fibres)
nylon	1.01–1.16
phenol-formaldehyde	1.27–1.30
	1.36–1.46 (filled with paper fibres)
polyester	1.38–1.41
polyethylene	0.91–0.95
polypropylene	0.85–0.92
polystyrene	1.04–1.08
polyurethane	2.0–2.2
poly (vinyl chloride)	1.38–1.41 (unplasticized)
	1.19–1.35 (plasticized)
silicone	1.65
urea-formaldehyde	1.50
	1.80–2.10 (filled with paper fibres)

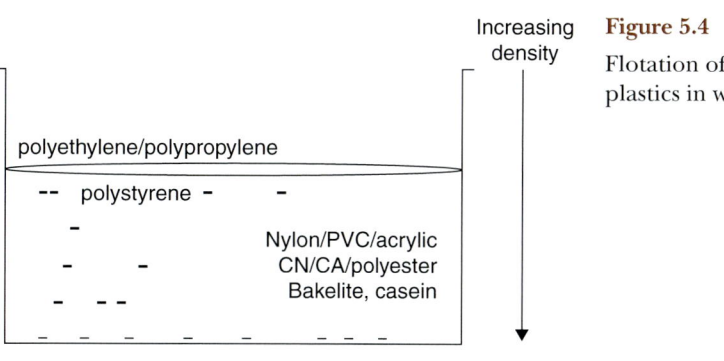

Increasing density

Figure 5.4

Flotation of various plastics in water.

fibres often present in formaldehyde plastics will affect flotation by absorbing water from the saturated solutions thus giving unreliable results (Coxon, 1993). Effects due to the surface tension of water or other fluids on the results may be minimized by adding a few drops of detergent (Williams et al., 1998).

5.2.2   Hardness

Hardness is another property which can be roughly tested with a fingernail applied to the underside of an object. Plastics which can be marked with a fingernail include polyethylene, polypropylene, plasticized PVC and polyurethane while other types are not affected. A rather unusual test is to tap the plastic firmly with a fingernail. If the sound is metallic, the plastic is likely to contain polystyrene.

5.2.3   Effect of heat

One of the most discussed tests for identifying plastics in museum and private collections is known as the hot pin test. It involves heating the point of a dressmaking pin in a flame until it glows and pushing it into the underside or hidden area of the unknown material. Whether the heated plastic softens or not identifies it as a thermoplastic or thermosetting material. Thermoplastics soften on heating while thermoplastics do not change state on heating until degradation temperature is attained. The hot pin test has been frequently used by private collectors to ascertain whether Bakelite is genuine or not. If a hot pin leaves no mark, the material may be Bakelite. The ability to mark the plastic with a hot pin is recorded together with odours detected.

Today, the application of a hot pin to plastics is no longer recommended by collectors and dealers as it was in the 1990s because of the risk of physical damage to the piece manifested by holes or burn marks. Such disfigurations result in a loss of commercial value (Parry, 1996). Instead of inserting hot pins, dealers hold the edge of the piece under hot running water for up to 30 seconds before examining and smelling it. The 'hot water test' is most frequently used to distinguish Bakelite from Fakelite because the latter does not develop Bakelite's characteristic odour of phenol. Because hot water cannot attain the high temperatures of heated metal pins (200–300°C), results from the two tests are poorly comparable.

Two types of heating test are used to identify plastics. The burn or flame test requires a sample of approximately $2 \times 4\,cm$ (Braun, 1996). A Bunsen burner is adjusted to its lowest setting and forceps are used to hold the sample over the flame. The colour of the flame and behaviour of the sample are recorded including its ability to melt or drip and whether it self-extinguishes or continues to burn after removal from the flame. Acrylonitrile-butadiene-styrene, acrylics, cellulose acetate, cellulose nitrate, polystyrene, polyurethane and polyesters continue to burn while nylon, polycarbonate and poly (vinyl chloride) are self-extinguishing. The flammability of plastics can be a useful tool in their identification, but the additives present affect their ability to burn. It should be noted that cellulose nitrate is highly flammable, especially when in good condition so a crumb-sized sample should be burnt first if its presence is suspected. Flame tests are illustrated in Figures 5.5 to 5.10. A summary of expected findings from the burn test is presented in Figures 5.11 and 5.12.

Figure 5.5
Flame test with cellulose nitrate. Burns explosively and sparks.

Any gas or vapour produced by burning is carefully smelled and recorded, bearing in mind that it may be toxic (polytetrafluoroethylene) or corrosive (cellulose acetate). The pH of any vapour generated is determined by holding Universal Indicator paper or pH indicator strips moistened with distilled water in its path. Plastics producing acidic vapours include cellulose nitrate and acetate, polyester, polyurethane, PVC. Those forming neutral vapours are polyethylene, polystyrene, acrylics, polycarbonate, silicones and epoxies. Alkaline products originate mainly from nylon and formaldehyde plastics.

The pyrolysis test involves heating a sample of plastic contained in a glass test tube (Braun, 1996). This test mainly provides information about the type of residue formed, if any, but all the measurements made during the burn test

Figure 5.6

Flame test with cellulose acetate. Burns slowly and steadily.

Figure 5.7

Flame test with low density polyethylene. Melts and drips during burning.

Figure 5.8
Flame test with
polystyrene. Burns with
sooty flame and choking
smell.

Figure 5.9
Flame test with nylon
6. Burns with frothing
and a blue-yellow
flame.

Figure 5.10

Flame test with melamine-formaldehyde. Expands and chars on burning.

Figure 5.11

Identification of plastics by heating-thermosetting materials.

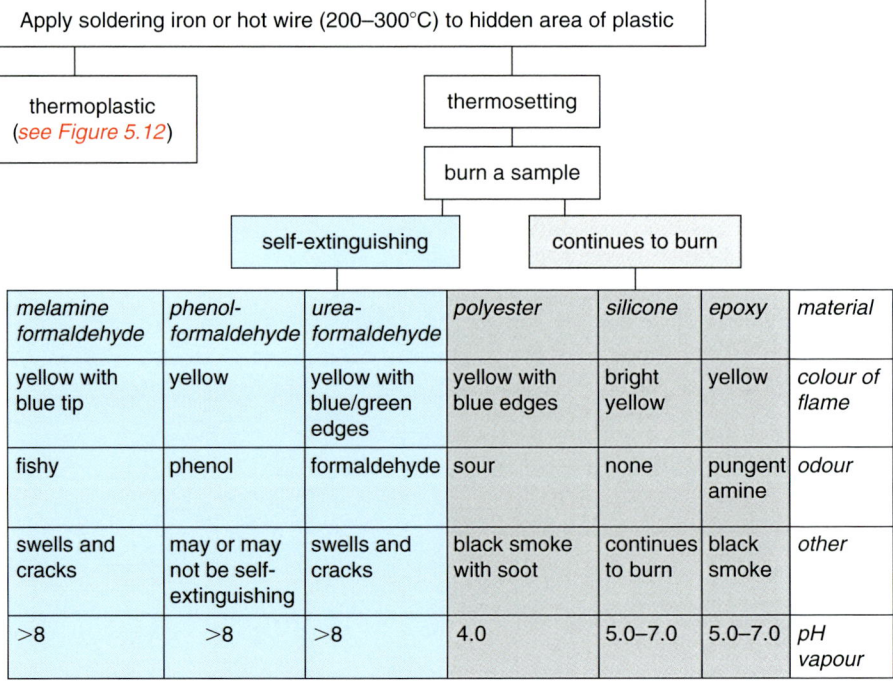

| Apply soldering iron or hot wire (200–300°C) to hidden area of plastic | | | | | | |

thermoplastic (*see Figure 5.12*)

thermosetting → burn a sample

self-extinguishing ‖ continues to burn

melamine formaldehyde	phenol-formaldehyde	urea-formaldehyde	polyester	silicone	epoxy	material
yellow with blue tip	yellow	yellow with blue/green edges	yellow with blue edges	bright yellow	yellow	*colour of flame*
fishy	phenol	formaldehyde	sour	none	pungent amine	*odour*
swells and cracks	may or may not be self-extinguishing	swells and cracks	black smoke with soot	continues to burn	black smoke	*other*
>8	>8	>8	4.0	5.0–7.0	5.0–7.0	*pH vapour*

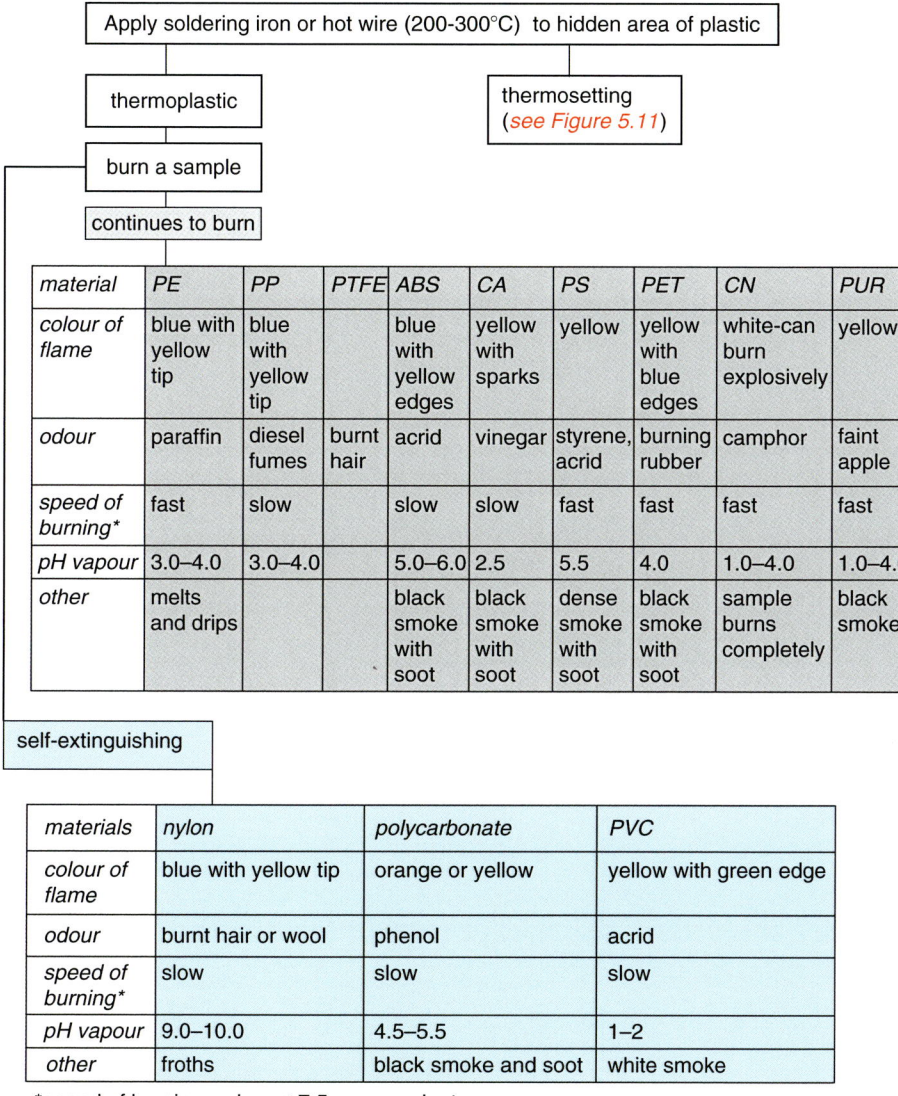

Figure 5.12

Identification of plastics by heating-thermoplastic materials.

material	PE	PP	PTFE	ABS	CA	PS	PET	CN	PUR
colour of flame	blue with yellow tip	blue with yellow tip		blue with yellow edges	yellow with sparks	yellow	yellow with blue edges	white-can burn explosively	yellow
odour	paraffin	diesel fumes	burnt hair	acrid	vinegar	styrene, acrid	burning rubber	camphor	faint apple
speed of burning*	fast	slow		slow	slow	fast	fast	fast	fast
pH vapour	3.0–4.0	3.0–4.0		5.0–6.0	2.5	5.5	4.0	1.0–4.0	1.0–4.0
other	melts and drips			black smoke with soot	black smoke with soot	dense smoke with soot	black smoke with soot	sample burns completely	black smoke

self-extinguishing

materials	nylon	polycarbonate	PVC
colour of flame	blue with yellow tip	orange or yellow	yellow with green edge
odour	burnt hair or wool	phenol	acrid
speed of burning*	slow	slow	slow
pH vapour	9.0–10.0	4.5–5.5	1–2
other	froths	black smoke and soot	white smoke

*speed of burning – slow < 7.5 cm per minute
fast > 7.5 cm per minute

can also be carried out, with the exception of the sample's ability to continue burning.

The Beilstein or copper wire test clearly indicates the presence of chlorine, bromine and iodine and was developed by Friedrich Konrad Beilstein (1838–1906) (Vogel, 1978). It is widely used to identify PVC. A clean, copper wire 30–40 cm long with a cork, or other heat-insulating material, at one end as an insulating handle, is heated with a Bunsen burner or other flame source until the flame is colourless to clean it of residual impurities and to develop a coating of copper (ll) oxide. The hot copper is then placed in contact with the plastic to be analysed so that a small piece melts onto it. The wire is returned to the base of the external part of the flame and the colour noted. A long-lasting green or blue-green flame caused by the formation of copper chloride denotes the presence of chlorine, mainly found in poly (vinyl chloride) and poly (vinylidene chloride), while other colours suggest the absence of chlorine (Figure 5.13). Although bromine and iodine can also produce a flame colour, fluorine is not detected because copper fluoride is not volatile.

Figure 5.13

Green flame produced by the Beilstein test indicates that the plastic contains chlorine and is usually poly (vinyl chloride).

The Beilstein test is very sensitive. Salt in sweat on plastics can also produce a green flame. Test results can also be confused by halogen-free materials. Some halogen-free compounds such as pyridine and quinoline derivatives, purines, acid amides and cyano compounds decompose in the flame to produce cyanides which also give a blue-green flame.

5.2.4   Solubility

A chemical is a solvent for another material if the molecules of the two are compatible and together produce a more stable system than the two materials separately. An understanding of polymer and solvent interactions can help to identify plastics because it gives information about their molecular weight and chemical properties including polarity, dispersion forces and ability to form hydrogen bonds. Higher molecular weight polymers and those with high crystallinity are less readily soluble than low molecular weight polymers with amorphous structures. Thermosets including unsaturated polyesters, epoxies, silicones, phenol-, urea- and melamine-formaldehydes and polyurethanes are poorly soluble due to their crosslinked, three-dimensional structures. Elastomers and synthetic rubbers are also crosslinked and insoluble, but swell on prolonged contact with most organic liquids (Vogel, 1978).

To examine solubility, solvent is placed in a glass vial, a small sample of plastic is added (for example a piece hole-punched from a film), the vial closed and the contents gently shaken. Vials are left to stand for 24 hours at ambient temperature and the contents examined. Changes in dimensions or swelling, texture or colour indicate interaction between the solvent and sample. If very little sample is available, the test can also be carried out on a micro scale by placing the unknown plastic on a glass microscope slide, which contains a well, adding solvent and closing with a glass cover slip. Any changes can be observed under a light microscope.

The most effective solvents for various polymers are shown in Table 5.4 (Verleye et al., 2001; Brydson, 1999). The presence of additives will affect

Table 5.4	Solubility of various polymers in organic solvents
Solvent	**Polymers dissolved at 20°C**
acetone	cellulose nitrate, cellulose acetates, poly(methyl methacrylate), polystyrene
ethanol	poly (vinyl acetate)
toluene xylene	polyethylene (warmed), polypropylene, polystyrene, poly (methyl methacrylate), ethylene-vinyl acetate copolymer, styrene-butadiene rubber
chloroform	cellulose nitrate, poly (vinylidene chloride), polycarbonate, poly (methyl methacrylate), polystyrene, polybutadiene, styrene-butadiene rubber
amyl acetate	polypropylene
diethyl ether	polystyrene
tetrahydrofuran (THF)	cellulose nitrate, cellulose acetate, poly (vinyl chloride), polystyrene, acrylonitrile-butadiene-styrene copolymer, styrene-butadiene rubber, acrylics
dimethylformamide (DMF)	cellulose nitrate, poly (ethylene terephthalate), polycarbonate, poly (vinyl chloride), acrylonitrile-butadiene-styrene copolymer

solubility of plastics. Inorganic additives such as fillers will not be soluble in the solvents effective for polymers whereas plasticizers are likely to exhibit similar solubility. Degradation of polymers will change their solubility compared with that when new. An increase in crosslinking due to degradation will increase the molecular weight thereby reducing the solubility, while the occurrence of chain scission reactions with time will have the opposite effect. Although water is not a solvent for synthetic polymers, it swells PVC and casein.

5.3   Spot tests for identifying polymers or their components

Spots tests involve dropping a reagent solution onto an unknown material and basing identification on the reaction observed. It is important to also apply the test to a reference material so that the reaction obtained with the unknown sample can be identified with confidence. In addition, blank spot tests (that is with no sample present) indicate the colour or other change due to the sample.

Spot tests and other wet chemical techniques have largely been superseded by instrumental analytical techniques, particularly infrared spectroscopy (Coxon, 1993). Today, the wider availability, lower costs and especially the non-destructive qualities of spectroscopy make it more attractive for identifying polymers than spot tests. Nonetheless, spot tests are still in use and can provide a rapid identification or confirmatory analysis when analytical instruments are not available. For that reason, a selection of the spot tests most frequently applied to polymers is presented here. Because these tests involve the use of toxic, corrosive or oxidizing chemicals, they should be performed in a fume cupboard and safety goggles, gloves and protective clothing should be worn while carrying them out.

Detailed and clear descriptions of the theory, procedures and health and safety aspects of the spot tests most relevant to materials present in museums and galleries may be found in the publication *Material Characterization Tests for Objects of Art and Archaeology* (Odegaard et al., 2000). Spot tests that conservators and conservation scientists have found most effective to analyse plastics are presented here.

5.3.1   Test for cellulose nitrate

The use of diphenylamine in concentrated sulphuric acid to identify cellulose nitrate was originally used in forensic investigations in the 1930s to detect the presence of explosives on suspects' hands (Cowan and Purdon, 1967). As a forensic test, wax casts were taken of the suspect's hands and the diphenylamine reagent applied to the wax. Sulphuric acid reacts with cellulose nitrate to form nitronium ions (NO_2^+) which then oxidize diphenylamine to form the dark blue dye, diphenylbenzidine violet.

The reagent is prepared by adding 90 mL concentrated sulphuric acid to 10 mL distilled water under stirring. Diphenylamine (5 g) is added slowly (Williams et al., 1998). There has been disagreement as to the concentrations of diphenylamine and sulphuric acid required and the most consistent results have been obtained with a 5 per cent solution of diphenylamine in 75–85 per cent sulphuric acid (Coxon, 1993). The reagent should either be stored in a dark glass bottle or in the dark as ultraviolet light discolours it. The test is highly sensitive and a single drop of the reagent in contact with cellulose nitrate will develop the blue colour immediately (Figure 5.14). A control test can be run simultaneously using a known sample of cellulose nitrate. The development of no colour or a colour other than dark blue is considered a negative result.

Figure 5.14
A single drop of diphenylamine reagent produces a dark blue colouration immediately when applied to a crumb of cellulose nitrate.

5.3.2 Test for cellulose acetate

Coxon (1993) has recommended a spot test to identify the presence of acetates present in cellulose acetate and poly (vinyl acetate). Four pre-mixed reagents in turn are applied to a sample of the unknown plastic. One millilitre of a 6 per cent solution by weight potassium hydroxide in methanol is added to the sample. Next, 1 to 2 drops of a saturated solution of hydroxylamine hydrochloride in methanol are added and the mixture is either left to stand for at least 3 minutes or warmed gently. One drop 1 per cent ferric chloride in water is added and the mixture shaken. Finally, a 10 per cent solution of hydrochloric acid is added dropwise with shaking until a colour develops. Burgundy red indicates the presence of cellulose acetate or poly (vinyl acetate). Pale purple-red colouration is said to indicate cyanoacrylate or cellulose nitrate, but the diphenylamine test should be carried out if cellulose nitrate is suspected.

5.3.3   Test to identify the presence of sulphur

Several wet chemical tests have been used to identify the presence of sulphur in plastics including rubbers (Verleye et al., 2001). A quick test to detect the presence of sulphur is based on a sodium azide reagent (Daniels and Ward, 1982). The test is based on the catalytic decomposition of a sodium azide/iodine solution by sulphur-containing groups resulting in the evolution of nitrogen bubbles:

$$2Na_2N_3 + I_2 \rightarrow 3N_2 + NaI$$

The major disadvantage of this test is the high toxicity of sodium azide and the fact that it can explode when heated (Sax, 1975). Sodium azide (3 g) is dissolved in 100 mL of 0.05 M iodine solution (20 g/L potassium iodide plus 12.7 g/L iodine). Three millilitres industrial methylated spirit is added as a wetting agent. The reagent should stand for at least 30 minutes prior to use and can be stored in a dark glass bottle for around three months. A small sample of plastic is finely divided, placed on a clean microscope slide and covered with a glass coverslip. A drop of sodium azide reagent is introduced at the edge of the coverslip and is drawn under by capillary action where it makes contact with the plastic. The sample is observed for one minute in transmitted light through a microscope set to deliver $\times 40$ magnification and an assessment made of the quantity of nitrogen bubbles produced. The quantity of sulphur present in the plastic is directly proportional to the concentration of bubbles produced by reaction with sodium azide.

5.3.4   Test for polyamides

A test to distinguish polyamides from other plastics is based on the reaction between gases evolved on pyrolysis and p-dimethylaminobenzaldehyde (also known as Ehrlich's reagent) in the presence of hydrochloric acid, which produces a red complex (Odegaard et al., 2000). The tip of a glass Pasteur pipette or capillary tube is sealed in a flame. The sample of plastic is pushed to the base of the glass. A thin strip of filter paper is inserted into the glass near the opening. Two drops of 14 per cent weight to volume p-dimethylaminobenzaldehyde solution (1.4 g p-dimethylaminobenzaldehyde diluted to 10 mL with ethanol) are applied to the filter paper followed by one drop concentrated hydrochloric acid. The open end of the glass is covered with laboratory wrapping film. The pipette is warmed and smoke generated by the plastic sample is allowed to come in contact with the saturated filter paper.

If the filter paper becomes red, the plastic is a polyamide such as nylon, or is protein-based, such as caein-formaldehyde. The development of a blue colouration indicates polycarbonate. No change in the colour of the filter paper suggests the plastic to be of another type.

5.4   Instrumental techniques to identify plastics

Simple tests can only indicate which polymer type the plastic contains. To identify materials more precisely, it is necessary to use instrumental analytical methods. Each technique provides specific information either about which polymers or which additives are present, so it is usually necessary to use several in combination. For example, gas chromatography-mass spectrometry (GC-MS) is a destructive technique which allows identification of the polymer, plasticizer and other organic components. X-Ray Fluorescence (XRF) spectroscopy is an effective, non-destructive surface technique for identifying inorganic fillers, pigments and metal components, but cannot be used to identify polymers.

Since the 1990s, many instrumental techniques have become cheaper due to the wider availability of technology, particularly of computers used to process the extensive data produced and to help with interpretation. Since most cultural institutions operate on a limited budget, they are unlikely to have regular access to the most recently developed analytical equipment, although increasing collaboration with universities is slowly changing this situation. In addition, non-destructive techniques are preferred over destructive ones and qualitative results are usually more relevant than quantitative determinations. In 2000, a European COST Action G8 group was established with a main objective of increasing the knowledge of museum objects by non-destructive analysis (COST Action G8, 2006). The action finished in 2006. Conferences organized by that group, including Conservation Science 2002, have highlighted those instrumental techniques which are of greatest importance to conservation professionals and in which experience has developed (Townsend et al., 2002).

The instrumental analytical techniques discussed in this chapter are those used most frequently to identify plastics in collections. There are many techniques that are used in the plastics industry or university research laboratory which provide extensive information about synthetic materials, but which have not yet found a place in the conservation workshop. This may be attributed to the high cost of the instruments and their maintenance. For this reason they have not been included here. High resolution solid state nuclear magnetic resonance spectroscopy is one such technique which may well be found in many museum laboratories by 2015, but is not available to such institutions today (Lambert et al., 2000). Descriptions of instrumental analytical techniques have been divided into those used to identify polymers, those to examine fillers and those to characterize plasticizers, stabilizers and flame retardants.

5.4.1   Instrumental techniques to identify polymeric component of plastics

The most frequently used technique to identify polymers is Fourier Transform Infrared spectroscopy. Due to developments in the 1990s, it can be used non-destructively and the expansion of electronic reference libraries has helped to interpret spectra. The addition of a pyrolysis unit has allowed small samples of polymers to be analysed by gas chromatography-mass chromatography. Polymers are high molecular weight materials and difficult to dissolve for presentation to analytical equipment. Thermal analysis has been used since the 1960s to analyse polymers and has been developed to require samples as small as 10 mg. The application of Raman spectroscopy in the analysis of polymers is growing rapidly and, as portable instruments become less expensive, this non-destructive technique is expected to be found in most conservation laboratories by 2010.

Fourier transform infrared (FTIR) spectroscopy
The most widely available technique for identifying polymer types is Fourier Transform Infrared (FTIR) spectroscopy. Samples are exposed to infrared radiation (4000–200 wavelengths per centimetre or cm^{-1}) causing chemical bonds to vibrate at specific frequencies, corresponding to particular energies. The stronger the vibrating atoms are bonded, the higher the wavenumber that absorption occurs at. The wavenumber also increases with reducing mass of vibrating atoms. A Fourier Transform Infrared spectrophotometer determines the absorption at all wavelengths simultaneously using a Michelson interferometer and plots the absorption intensity for each wavenumber as a graph or spectrum. Spectra are traditionally displayed showing transmittance (peaks pointing to the bottom of the page) instead for absorption (peaks pointing upwards).

There are several techniques for preparing a sample of plastic for infrared spectroscopy and selection depends on the mechanical properties of the sample. The traditional preparation technique for crystalline materials, in which a sample is finely ground into potassium bromide (KBr) powder before pressing into a tablet one centimetre in diameter or a smaller micropellet, is rarely effective for plastics. Thermoplastics tend to soften in the heat generated by grinding and adhere to the KBr granules, making it difficult to disperse them evenly. Thermosets tend to be brittle and break into unevenly sized pieces thus preventing homogeneous mixing with KBr. If the infrared beam makes contact with a large particle of plastic, it is likely to be scattered or reflected from surfaces rather than absorbed, resulting in poor quality spectra. If the polymer component can be dissolved in an appropriate solvent, the solution can either be mixed with the KBr, the two ground together until the solvent has evaporated and a tablet pressed or, alternatively, a few drops of solution can be applied to a pressed tablet and allowed to dry.

If the plastic takes the form of a thin film, it can be exposed to the infrared radiation without further preparation. However, if the film is thicker than around 0.5–0.75 mm, too much of the infrared radiation is absorbed to produce a well-resolved spectrum.

In the late 1990s, an accessory was developed which enabled non-destructive examination of surfaces and so was ideal for analysis of plastics in museum collections. Attenuated Total Reflection FTIR (ATR-FTIR) spectroscopy requires samples to be placed on a diamond crystal with a diameter of 2 mm, through which the infrared beam is reflected (Figure 5.15). The quality of spectra is dependent on intimate contact between the diamond crystal and the sample, so a pressure device is used to achieve this (Figure 5.16). Diamond has a higher refractive index than most plastics (2.4 and approximately 1.5 respectively) so when the beam comes into contact with the sample, it penetrates only about 2 µm into the material before being totally internally reflected by it – the resulting beam is then directed into the detector. The detector examines the frequencies absorbed compared with those present in the incident radiation and produces a spectrum or trace which is definitive for all infrared-sensitive chemical bonds present in a particular material or mixture. A spectrum can be obtained 30 seconds after placing the sample in the instrument.

Figure 5.15
Attenuated Total Reflection accessory for FTIR spectrometer with a diamond crystal in the centre through which the infrared beam is reflected.

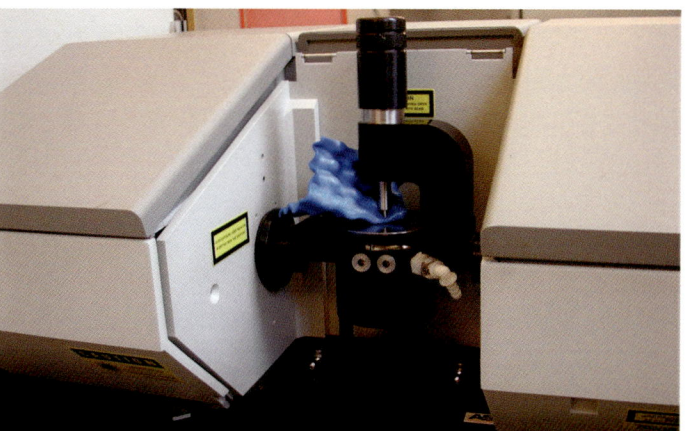

Figure 5.16
Plastic samples must make good contact with the surface of the diamond to obtain well-resolved spectra. This is achieved using a pressure device.

Unknown plastics may be identified either by comparing spectra with those from reference or known samples, with commercial reference libraries such as those specific to art, architecture and archaeological materials produced by the Infrared and Raman Users Group (IRUG, 2007) or by using tables to correlate infrared absorption energies with specific chemical bonds. Spectra of polymers obtained using an ATR accessory and those from a KBr tablet look similar and can be compared directly. However, when comparing spectra obtained using an ATR accessory with those obtained by KBr, it should be considered that the positions of wavelengths may be shifted by $5\,\mathrm{cm}^{-1}$. In addition, the relative intensities of bands present in ATR spectra can vary from those seen in conventional transmission spectra.

ATR-FTIR transmission spectra of the plastics most frequently encountered in museum collections, together with assignments for the characteristic peaks for each material, are assembled in Appendix 2. Unless a plastic sample is prepared by extracting in solvent or pyrolyzed to separate the polymer from inorganic additives prior to analysis by spectroscopy, its FTIR spectra shows absorbances by all components present in the plastic including additives, fillers and colouring materials, so can be complicated to interpret (Figure 5.17 and Table 5.5). Because the absorption bands attributed to inorganic fillers in plastics occur around $400\,\mathrm{cm}^{-1}$, $1000\,\mathrm{cm}^{-1}$ and $700\,\mathrm{cm}^{-1}$, they often overlap with C-H stretches and other bands due to organic materials in the polymer component.

Figure 5.17

ATR-FTIR spectrum of poly (vinyl chloride) photograph pocket with phthalate plasticizer shows peaks due to both PVC and plasticizer. Units are wavelengths cm^{-1} on the horizontal axis and per cent transmittance on the vertical axis.

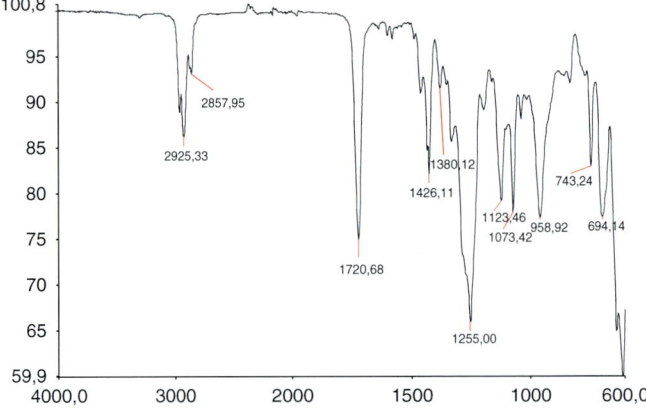

Raman spectroscopy

Raman spectroscopy is a non-destructive, non-contact technique which is increasingly being applied to identify polymers. A laser is directed onto the surface of the material to be analysed. Most of the incident radiation is reflected unchanged from the surface. However, a small proportion interacts with the molecules in the material and is scattered. The scattered portion of light,

Wavenumber (cm⁻¹)	Type of vibration	Assignment
Table 5.5 Peak table for ATR-FTIR spectrum of plasticized PVC. (Peaks attributed to plasticizer are shown in italics)		
3000–2840	stretching	CH_2 and CH
1720	*stretching*	*C=O*
1464	*stretching*	*CH*
1426	in-plane deformation	CH_2
1380	*stretching*	*CH*
1332	wagging vibration	CH
1258	*stretching*	*C—O*
1252	in-plane deformation	CH
1123	*stretching*	*C—O*
1073	*stretching*	*C—O*
958	*stretching*	*C—C*
876	rocking vibration	CH_2
743	*stretching*	*CH*
615	stretching	C—Cl

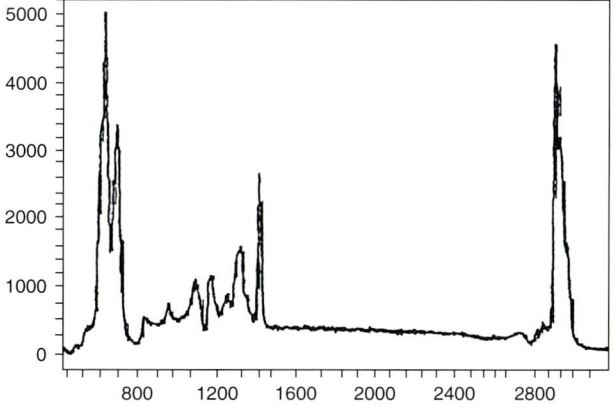

Figure 5.18
Raman spectrum of poly (vinyl chloride). Units are wavelengths cm⁻¹ on the horizontal axis and energy on the vertical axis.

known as the Raman effect, is collected to produce a spectrum. Each material has a unique spectrum associated with it and therefore each one acts as a fingerprint with which to identify materials (Forrest, 2002) (Figure 5.18). Because of the way in which the Raman effect is produced, the information obtained from infrared spectroscopy can also be found in Raman spectra. Some vibrational modes for polymers appear only in the infrared spectrum and others are only present in the Raman spectrum. In general, the more symmetric the molecule, the greater the differences between the infrared and Raman spectra. Raman spectra contain much information in the lower frequency range where there is little absorbance in infrared spectra. Bands due to C-C dominate the Raman spectrum while C=O and C—O are the most intense in the infrared spectrum (Koenig, 1999).

The Raman effect is very small and until recently its use as an analytical tool was limited by a lack of suitable equipment. Recent advances, particularly the development of higher resolution and lower cost systems, mean that Raman spectroscopy is now used to analyse a wide range of materials. Raman spectroscopy is totally non-destructive although dark samples may become very hot and burn. Using a microscope, spectra can be recorded from very small samples, as small as 1–2 μm across, and from the different layers present within a laminate. Whereas ATR-FTIR gives information from a depth up to around 2 μm into the material, Raman spectroscopy has a penetration depth in the order of millimetres.

Pyrolysis gas chromatography-mass spectrometry (py-GC-MS)

Py-GC-MS can also be used to identify the polymer type in a plastic (Learner, 2001). The principle of the technique is based on the use of heat to volatilize and break down macromolecules into smaller components capable of being analysed using GC-MS (Table 5.6). Degradation mechanisms for polymers under pyrolysis are free radical processes initiated by bond dissociation due to the heat. The specific pathway followed by a particular polymer is dependant on the strength of the polymer bonds and the structure of the polymer chain. Pathways may be described as random scission, unzipping and side group elimination. Polyolefins such as polyethylene and polypropylene follow the random scission pathway and break into pieces of the original molecule to form

Table 5.6 Characteristic species identified in pyrograms of polymers	
Polymer	**Species produced on pyrolysis**
polyethylene	straight chain hydrocarbons: n-hexane, n-pentane
poly (methyl methacrylate)	monomers: methyl methacrylate
polypropylene	branched chain hydrocarbons: 2,4-dimethyl heptane
polystyrene	benzene, toluene, ethylbenzene, styrene
poly (vinyl chloride)	side group products: hydrogen chloride, benzene, toluene, chlorobenzenes

oligomers and hydrocarbons. Poly (methyl methacrylate) and other acrylics unzip at pyrolysis temperatures and produce monomers. Chlorine-containing polymers such as PVC break at the weakest bonds under pyrolysis (the side chains) and form hydrogen chloride, benzene, toluene and other aromatics.

The sample (ca. $100\,\mu g$) can either be extracted in solvent before presentation to the instrument or can be introduced without preparation. Samples are gradually heated up to around $750°C$ and the resulting volatile products are carried onto the gas chromatography column by an inert carrier gas. Separation occurs as the components partition themselves between the stationary phase on the inner wall of the column and the mobile phase (the carrier gas). The time taken for a given molecule to traverse the entire length of the column is known as the retention time. The retention time is a function of the chemical structure of the component, the column type and the temperature profile it has been subjected to. It depends on the relative affinity of the compound for the stationary and mobile phases.

A mass spectrometer then detects the components that elute from the end of the gas chromatography column. In the mass spectrometer, energetic electrons bombard the component molecules, ionizing some of them. This ionization process can also produce fragment ions which often provide structural information about the molecule. The ions are then accelerated by an electric field and enter a mass analyser, where they are separated according to their mass-to-charge ratios. By plotting the abundance of ions as a function of mass-to-charge ratio, a mass spectrum is generated. The mass spectrum can be a unique fingerprint, allowing identification of unknown compounds. This fingerprint is compared with a database of over $107\,000$ unique chemical compounds (NIST, 2006).

Differential scanning calorimetry (DSC)

Thermal analysis involves examining the change in a property of a material on heating or cooling. The property may change in weight or mechanical properties. DSC is used extensively to identify polymers and polymer blends (Forrest, 2002). It differentiates between various materials from their melting points (Tm), glass transition temperatures or Tg (temperature at which polymers change from a glassy state to a rubbery state) and the appearance of their melting endotherms (the heat flow required to raise the temperature of the material) (Table 5.7). A small sample (around $2\,mg$) is heated in a pan and the heat required to increase the temperature is compared with the heat required to increase the temperature of an identical, empty pan.

Thermogravimetric analysis (TGA)

TGA examines change in weight during heating in nitrogen or air. It allows quantitative measurements of the weight changes due to volatilization or other

Table 5.7	Glass transition temperature (Tg) and melting temperatures (Tm) for crystalline polymers	
Polymer	Tg (°C)	Tm (°C)
nylon 6	50	215
nylon 66	60	266
polycarbonate	150	225
polyethylene – low density	−20	105
polyethylene – high density	−20	135
poly (ethylene terephthalate)	67	256
polystyrene-atactic	95	–
polystyrene-isotactic	100	230
polytetrafluroethylene	115	327

phase changes of the polymer, plasticizer and inorganic additives in plastics. The temperature at which weight changes take place is the information used for identification. TGA requires small samples, around 10 mg, and short heating periods of around 60 minutes. Weight loss derivative is often plotted against temperature, which highlights the presence of a polymer blend or the presence of fillers and additives. The temperature at which polymers lose 50 per cent of their initial weight can be used to identify them (Table 5.8). The thermal analyser may also be connected to a mass spectrometer (TGA-MS) or infrared spectrometer to identify the products of thermal degradation.

Table 5.8 Thermal decomposition temperatures for pure polymers	
Polymer type	Temperature at which 50% weight loss occurs* (°C)
polyethylene	414
poly (methyl methacrylate)	327
polypropylene	387
polytetrafluoroethylene	509
poly (vinyl chloride)	260
*temperature at which the polymer loses 50% weight heating in vacuum for 30 minutes after 5 minutes preheating	

5.4.2 Instrumental techniques to identify the filler component of plastics

Fillers are used to add bulk, increase opacity and to improve tensile strength of polymers. They usually comprise inorganic materials in particle or fibre form although organic materials including wood flour, nylon and polypropylene fibres may also be present. Inorganic fillers include calcium carbonate, magnesium carbonate and barium sulphate. More details of filler types and their functions can be found in Chapter 3. A destructive technique used to identify inorganic fillers is to remove the polymer and other organic components prior to examination. Samples are subjected to furnace ashing involving heating at 500–600°C (ISO 345-1, 1997). Glass fibres begin to melt around 700°C, so ashing above this temperature is not recommended.

Energy dispersive X-ray analysis (EDAX, EDX or EDS)

A non-destructive technique to identify fillers qualitatively is Energy Dispersive X-ray analysis. EDAX is used to identify the elemental composition of a material and works as an integrated feature of a scanning electron microscope (SEM). The sample is bombarded with an electron beam inside the scanning electron microscope. At rest, an atom within the sample contains ground state or unexited electrons situated in concentric shells around the nucleus. Bombarding it with the electron beam excites an electron in an inner shell, causing its ejection and resulting in the formation of an electron hole within the atom's electronic structure. An electron from an outer, higher-energy shell then fills the hole, and the excess energy of that electron is released in the form of an X-ray.

Measuring the amounts of energy present in the X-rays being released by a specimen reveals the identity of the atom from which the X-ray was emitted. The EDAX spectrum is a plot of how frequently an X-ray is received for each energy level. An EDAX spectrum normally displays peaks corresponding to the energy levels for which the most X-rays had been received. Each of these peaks is unique to an atom, and therefore corresponds to a single element, which results in identification of fillers. The principles and techniques of scanning electron microscopy and elemental analysis are described by Goldstein (1981). Electrons penetrate to a depth of 2–3 μm and EDAX can be used to scan plastics surfaces for elements. As a result, unless a cross-section or sample is taken, or the filler is separated from the polymer by ashing, only the fillers close to the surfaces can be analysed, and their concentration may not represent that of the bulk.

X-ray fluorescence (XRF)

X-ray fluorescence is a non-destructive technique which detects elements using the same principles as EDAX. Samples are exposed to short wavelength X-rays or gamma rays, which cause ejection of one or more electrons from the inner orbital of the atom and the formation of a hole. The atom becomes unstable and electrons from higher orbitals fall into a lower orbital to fill the hole. In falling, energy is released as a photon and the material fluoresces radiation with energy equal to the energy difference between the two orbitals, which is characteristic to the atom. In principle, the lightest element that can be analysed is beryllium, but due to instrumental limitations and low X-ray yields for light elements, it is often difficult to quantify elements lighter than sodium, unless background corrections and very comprehensive corrections for elements are made. Details of the theory and practice of XRF in art and archaeology can be found in a review article by Mantler and Schreiner (Mantler and Schreiner, 2000).

X-ray fluorescence is used to detect elemental magnesium, aluminium, silicon and iron present in glass fibres. They exhibit detection limits around 5 ppm. Handheld XRF analysers are now available and finding increasing use in museums because they can be used in storage areas and displays to examine objects without the inconvenience of transportation.

Fourier transform infrared (FTIR) spectroscopy

Inorganic ash from burning polymers to separate the organic polymer from fillers can be identified by infrared spectroscopy. The technique of infrared spectroscopy has been described previously in this chapter. The bands characteristic to inorganic fillers are shown in Table 5.9.

Table 5.9 Characteristic FTIR absorption bands for inorganic fillers in plastics		
Filler	**Characteristic bands (cm^{-1})**	**Comments**
silica	950–1330 maximum at 1050–1100	broad band
silicates	850–1300 maximum at 950–1100	broad band
calcium carbonate	1420, 870, 710	
barium sulphate	1080, 610	

5.4.3   Instrumental techniques to identify plasticizer

Plasticizers are attached to the polymer by weak physical bonds rather than chemical ones, so can be separated readily by solvent extraction. A suitable solvent should selectively dissolve the plasticizer from the polymer. Methanol is frequently employed. Excess methanol should be dried off at 105°C. Infrared spectroscopy can be used to identify plasticizers by smearing the sample on one side of a KBr tablet and running a transmission spectrum or by examining the neat extract by ATR-FTIR spectroscopy. The resulting spectra can be compared with those in a specialized database of additives (Scholl, 1981).

Gas chromatography-mass spectrometry (GC-MS)

Gas chromatography-mass spectrometry is a destructive technique which is used to identify plasticizers. The gas chromatograph (GC) portion separates the chemical mixture into smaller molecular weight components, which are swept into the mass spectrometer (MS) to be identified and quantified. The GC separates chemicals based on their volatility. In general, small molecules travel more quickly than larger molecules. The MS is used to identify chemicals based on their structure. Methodology for GC-MS is detailed by Skoog et al. (1998).

Samples are generally extracted with methylene chloride, the extract concentrated, and the resulting extract analysed by GC-MS. This technique is frequently used to quantify plasticizer in consumer products. For example, analysis of chemical substances in plastic toys for pets by the Danish Environmental Protection Agency involved extracting 50 mg of finely divided toy in dimethylene chloride at ambient temperature. Any dissolved PVC was precipitated by adding methanol. The extract was centrifuged and analysed by GC-MS. A phthalate plasticizer content of 11–54 per cent by weight was determined (Danish Environmental Protection Agency, 2005).

Detection limits for plasticizers by GC-MS range from parts per million (ppm) to parts per billion (ppb). A typical chromatogram of di-ethylhexyl phthalate (DEHP), the most commonly used plasticizer for PVC, is shown in Figure 5.19. The mass spectrum shown is a molecular fragmentation pattern which identifies this peak as DEHP. The size of the peak in the chromatogram relates to the amount in the sample. Pyrolysis gas chromatography-mass spectrometry can also be used to identify plasticizers if sampling of the plastics object is permitted. A sample of between 0.1 and 5.0 mg is necessary.

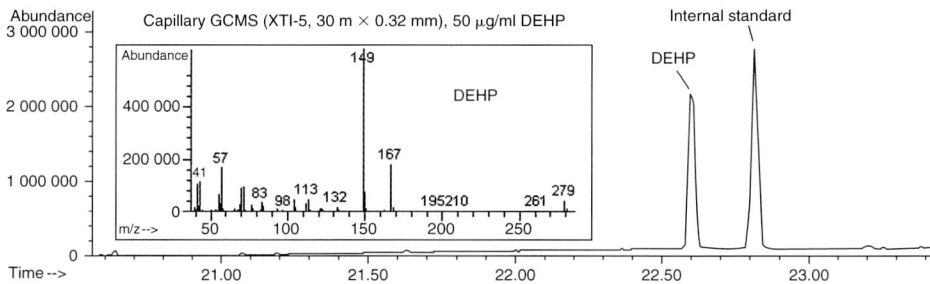

Figure 5.19

A chromatogram characteristic of di-ethylhexyl phthalate (DEHP), the most commonly used plasticizer for PVC. The mass spectrum on the right-hand side is a molecular fragmentation pattern which identifies this peak as DEHP.

GC-MS can also be used to identify the volatile and semi-volatile components of plasticizer via headspace analysis (the analysis of gases evolved from a solid or liquid). The technique is non-destructive if the object is small enough to be enclosed in an airtight bag or glass container fitted with a Teflon septum in the cap. There are three approaches to collecting gases evolved from the plastic. Glass tubes containing adsorbent can be placed in the bags prior to sealing or gases may be collected from the bag or container by syringe. A third technique, which was developed in the 1990s and is assuming increasing importance in the study of volatile organic compounds (VOC), is solid phase microextraction (SPME). An SPME needle may be compared with a very short gas chromatography column turned inside out. It has an outer polymer coating which selectively adsorbs volatile materials when exposed to air in a container. Volatiles are then desorbed in the hot inlet to the gas chromatograph and subsequently identified using a mass spectrometer (Larroque et al., 2006).

The enclosed plastic object may be warmed gently in an oven to promote evaporation of plasticizer if it is desirable to reduce the time required for measurement. After a collection time ranging from 15 minutes to 3 hours, depending on the volatility of the plasticizers, gases from the adsorbent are thermally desorbed and analysed by GC-MS, while gases collected by syringe are injected directly into a gas chromatograph. Components are identified either by comparison with standard samples or by comparing the respective mass spectra with mass spectra from reference spectral libraries such as NIST (National Institute of Science and Technology) (NIST, 2006) (Figure 5.20).

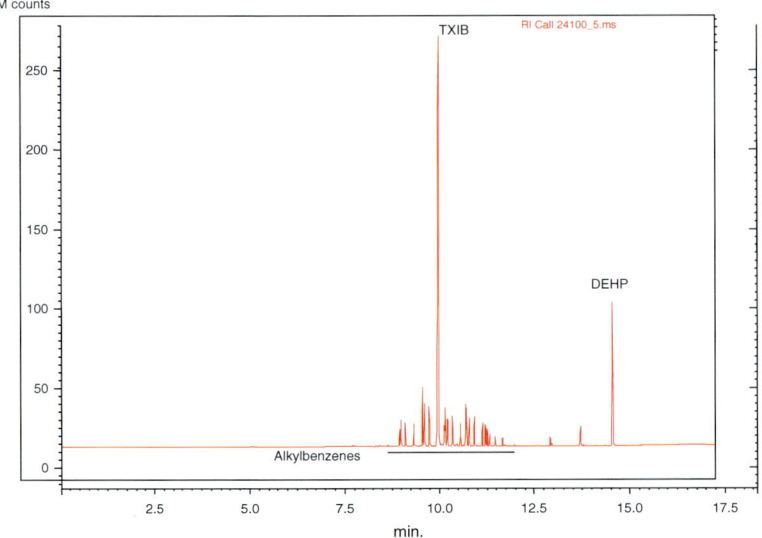

Figure 5.20

Gas chromatogram of vapour above plasticized PVC flooring collected by headspace analysis after warming to 70°C. The chromatogram shows the presence of two plasticizers: DEHP and TXIB (2,2,4-trimethyl-1,3-pentanediol diisobutyrate). TXIB imparts lower tack and increased stain resistance to PVC products compared with those containing DEHP alone.

Energy dispersive X-ray spectroscopy (EDAX)

Once a plasticizer has been identified, EDAX can be used to examine its distribution in the plastic matrix if the two components contain at least one element which is not common to both. Examination can be non-destructive if a low vacuum scanning electron microscope is used because the technique does not require the application of a conductive coating. The plastic surface is mapped for elements.

The PVC life support hoses from the suits used in the Apollo Space mission dating from the late 1960s were examined for distribution of plasticizer (Shashoua, 2001). PVC tubing plasticized with phthalates was used to carry water and oxygen from portable life support systems to astronauts during lunar activity. This grade of PVC tubing was manufactured to transport food and drink in the 1960s. Today, the tubing is highly degraded and products of hydrolysis of the phthalate plasticizer, mainly phthalic acid, is present as white crystals at the surfaces of discoloured PVC. Phthalic acid contains the elements carbon, hydrogen and oxygen, while PVC contains carbon, hydrogen and chlorine. The presence of chlorine is, therefore, indicative only of PVC polymer while the presence of oxygen is indicative only of plasticizer and its degradation products. Using EDAX to map for carbon, chlorine and oxygen revealed the location of crystalline phthalic acid at surfaces of life support hoses (Figures 5.21–5.23).

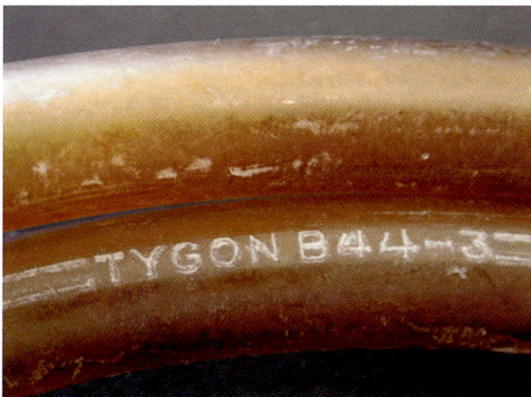

Figure 5.21
Yellowed life support tubing from Apollo space suit showing crystalline phthalic acid on outer surfaces. The tube is 15 mm in diameter.

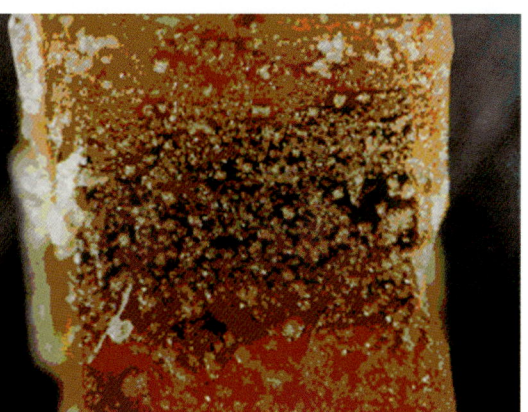

Figure 5.22
Crystals on inside surface of life support tube.

Figure 5.23

Low vacuum scanning electron microscope scan (far left) shows crystalline, leaf-shaped, degradation products on inside surfaces of life support tube. Elemental maps for chlorine (centre) and oxygen (far right) show the crystals contain higher concentrations of oxygen (brighter) than the tube, indicating the presence of phthalic acid. Sample size 1 mm × 1 mm.

5.4.4   Instrumental techniques to identify stabilizers and flame retardant components of plastics

Stabilizers are included in all plastics formulations to stop or slow the loss of physical properties, which occurs when polymers react with ultraviolet radiation, heat, oxygen and ozone. Stabilizers are not usually volatile because they are designed to remain in the plastic for its lifetime and are present in small quantities comprising up to 0.5 per cent of the weight of the plastic, so it is necessary to separate them from the rest of the formulation in order to identify them.

High performance liquid chromatography (HPLC)

A wide range of stabilizers in plastics is identified commercially using high performance liquid chromatography, a separation technique based on the distribution of compounds between two phases known as the stationary phase and mobile phase. The stationary phase comprises a thin layer created on the surface of fine particles and the mobile phase comprises the liquid flowing over the particles. Each component in a sample has a different distribution equilibrium depending on its solubility in the phases and/or molecular size. As a result, the components move at different speeds over the stationary phase and are thereby separated from each other. The sample is dissolved using a suitable solvent, a non-solvent such as methanol is added, and the extract presented to the HPLC.

The column is a stainless steel (or polymeric) tube which is packed with spherical solid particles. Blends of methanol and water or ethyl acetate and acetonitrile, comprising the mobile phase, are constantly fed into the column inlet at a constant rate by a liquid pump. A sample is injected from a sample injector, located near the column inlet. The injected sample enters the column with the mobile phase and the components in the sample migrate through it, passing between the stationary and mobile phases. Compounds that tend to be distributed in the mobile phase therefore migrate faster through the column while compounds that tend to be distributed in the stationary phase migrate slower. In this way, each component is separated on the column and sequentially elutes from the outlet. Each compound eluting from the column is monitored by a detector connected to its outlet, based on the response to UV light. A chromatogram shows the time required for a compound to elute (called retention time) and its concentration. Retention time is used to identify the sample and peak size to quantify components by comparing with standard reference materials. Antioxidants that can be identified using HPLC include hindered phenolics, butylated hydroxytoluene (BHT) and triclosan (5-chloro-2-[2,4-dichlorophenoxy]phenol) (Forrest, 2002).

Thin layer chromatography (TLC)

A simpler, lower cost technique which can readily separate the antioxidants from plastic extracts and give a qualitative analysis is thin layer chromatography (BS6630, 1985). In thin layer chromatography (TLC), the stationary phase is comprised of a thin layer of adsorbent such as cellulose, alumina, or silica gel on a plastic sheet, thick aluminium foil, or a glass plate. A small spot of solution containing the sample is applied to a plate, about 1 cm from the base. The plate is then placed in a sealed container which holds a suitable solvent, such as ethanol, so that it does not come into contact with the spots. The solvent moves up the plate by capillary action and meets the sample mixture, which is dissolved and is carried up the plate by the solvent. Components in the sample mixture travel at different rates due to differences in solubility in the solvent, and due to differences in their attraction to the plate.

When the solvent has reached the top of the plate, the plate is removed from the developing chamber, dried, and the separated components of the mixture are examined. Because the components are usually colourless or white, a UV lamp or blacklight (UV_{254}) is used to make the spots fluoresce. Iodine vapour, copper sulphate and other coloured reagents can also be applied to induce characteristic colours to spots of antioxidants. Once visible, the R_f values of the spots can be determined by dividing the distance travelled by the coloured product by the total distance travelled by the solvent (the solvent front). Each antioxidant has a distinctive R_f value. These values should be the same regardless of the distance travelled by the solvent, but are dependant on the solvent used, and the type of TLC plate. For this reason, TLC should be performed on reference samples of antioxidants at the same time as the unknown materials are analysed.

Electron dispersive X-ray spectroscopy and X-ray fluoresence

Flame retardants and heat stabilizers contain halogens, particularly bromine or phosphorus, or metal soaps such as stearates, palmitates and octoates of cadmium, barium, calcium and zinc, and salts of heavy metals. The halogens and metallic elements of stabilizers and flame retardants can be identified non-destructively within the plastic matrix using EDAX or XRF as described earlier in this chapter for identification of fillers. XRF can detect all elements heavier than sodium and some at concentrations less than 10 ppm. EDAX has been used to screen the plastic housings of household appliances, including coffee machines, irons, toasters and fan heaters, for brominated flame retardants and concentrations ranging from 0.02 to 9.6 per cent by weight were measured (Danish Environmental Protection Agency, 2001). The detection limit for bromine and antimony has been established as 0.01 per cent by weight.

References

Braun, D. (1996). *Simple methods for the identification of plastics*, 3rd edition. Hanser/Gardner Publishing Inc., Cincinnati.

Brydson, J. A. (1999). *Plastics Materials*, 6th edition. Butterworth-Heinemann, Oxford.

British Standards (1985). BS6630:1985: Method for identification of antidegradants in rubber and rubber products by thin layer chromatography. British Standards Institute, London.

Cloutier, H. and Prud'homme, R. E. (1993). Rapid identification of thermoplastic polymers. *Journal of Chemical Education*, **62** (9), 815–819.

Cost Action G8. (2006). *Non-destructive analysis and testing of museum objects* [online]. Available from: http://www.srs.dl.ac.uk/arch/COST-G8/ [Accessed August 2006].

Cowan, M. E. and Purdon, P. L. (1967). A study of the paraffin test. *Journal of Forensic Science*, **12** (1), 15–19.

Coxon, H. C. (1993). Practical pitfalls in the identification of plastics. In *Postprints of Saving the Twentieth Century: The Conservation of Modern Materials* Ottawa, 15–20 September 1991 (D. Grattan, ed.) pp. 395–406. Canadian Conservation Institute.

Daniels, V. and Ward, S. (1982). A rapid test for the detection of substances which will tarnish silver. *Studies in Conservation*, **27**, 58–60.

Danish Environmental Protection Agency.(2001). *Analysis of bromine in electric parts. Arbejdsrapport 26/2001 Survey of chemical substances in toys for animals* [online in Danish]. Available from: http://www.mst.dk/udgiv/publikationer/2001/87-7944-835-6/html/helepubl.html [Accessed 20 August 2006].

Danish Environmental Protection Agency (2005). Survey of Chemical Substances in Consumer Products, *Survey of chemical substances in toys for animals* [online]. Available from: http://www.mst.dk/udgiv/publications/2005/87-7614-662/html/kap03_eng.html [Accessed 20 August 2006].

Davidov, C. and Dawes, G. R. (1988). *The Bakelite Jewelry Book*. Abbeville Press Inc., USA.

Forrest, M. J. (2002). Analysis of Plastics. *Rapra Review Report*, **13** (5). Rapra, Shawbury.

Health and Safety Executive (2007). *Control of Substances Hazardous to Health (CoSHH)* [online]. Available from: http://www.hse.gov.uk/coshh/ [Accessed 20 August 2007].

Garcia Fernandez-Villa, S. and San Andres Moya, M. (2005). Original patents as an aid to the study of the history and composition of semisynthetic plastics. *JAIC*, **44**, 95–102.

Goldstein, J. I. et al. (1981). *Scanning electron microscopy and x-ray microanalysis: a text for biologists, materials scientists, and geologists*. Plenum, New York.

Green, L. R. and Bradley, S. M. (1988). Investigation into the degradation and stabilization of the nitrocellulose in the museum collections. In *Preprints of Modern Organic Materials* Edinburgh, 14–15 April 1988 (M. Wright, ed.) pp. 81–95, SSCR.

IRUG (2007). *Infrared and Raman Users Group* [online]. Available from: http://www.irug.org [Accessed 20 August 2007].

ISO (1997). ISO 345-1:1997: Furnace Ashing.

Koenig, J. L. (1999). *Spectroscopy of Polymers*, 2nd edition. Elsevier, Amsterdam.

Lambert, J. B., Shawl, C. E. and Stearns, J. A. (2000). Nuclear magnetic resonance in archaeology. *Chemistry Society Review*, **29**, 175–182.

Larroque, V., Desauziers, V. and Mocho, P. (2006). Development of a solid phase micro-extraction (SPME) method for the sampling of VOC traces in indoor air. *Journal of Environmental Monitoring*, Jan., **8** (1), 106–111.

Learner, T. (2001). The analysis of synthetic paints by pyrolysis-gas chromatography-mass spectrometry (Py-GC-MS). *Studies in Conservation*, **46**, 225–241.

Mantler, M. and Schreiner, M. (2000). X-ray fluorescence spectrometry in art and archaeology. *X-ray Spectrometry*, **29**, 3–17.

NIST Scientific and Technical databases (2006). *Database of gas chromatograms* [online]. Available from: http://www.nist.gov/srd/nist1a.html [Accessed 15 August 2006].

Odegaard, N., Carroll, S. and Zimmt, W. S. (2000). *Material characterization tests for objects of art and archaeology*. Archetype publications, London.

Oxford Physical and Theoretical Chemistry Laboratory (2006). *Maximum Exposure Limits* [online]. Available from: http://physchem.ox.ac.uk/MSDS/mels.html [Accessed 20 August 2007].

Parry, K. (1996). *Karima Parry's Plastic Fantastic* [online]. Available from: http://www.plasticfantastic.com/testing.html [Accessed 21 July 2006].

Sax, N. I. (1975). *Dangerous properties of industrial materials*. Van Nostrand Reinhold, New York.

Scholl, F. (1981). *Atlas of Polymer and Plastics Analysis, Volume 3: Additives and processing aids: Spectra and Methods of Identification*, 2nd edition. Verlag Chemie GmbH, Weinheim.

Shashoua, Y. R. (2001). Inhibiting the degradation of plasticized poly(vinyl chloride) – a museum perspective. Ph.D. thesis, Danish Polymer Centre, Technical University of Denmark.

Skoog, A. et al. (1998). *Principles of instrumental analysis*, 5th edition. Saunders College Publishing, Philidelphia.

Texloc Closet (1997). *Plastic Materials–Identification Chart* [online]. Available from: http://www.texloc.com/closet/cl_plasticsid.html [Accessed 24 July 2006].

Townsend, J. H., Eremin, K. and Adriaens, A. (eds). (2002). *Conservation Science 2002*, Edinburgh, Scotland 22–24 May 2002. Archetype Publications, London.

Verleye, G. A. L., Roeges, N. P. G. and De Moor, M. O. (2001). *Easy identification of plastics and rubbers*. RAPRA Technology Ltd, Shawbury, United Kingdom.

Vogel, A. I. (1978). *Textbook of practical organic chemistry – qualitative organic analysis*, 4th edition. Longman, London.

Williams, R. S., Brooks, A. T., Williams, S. L., and Hinrichs, R. L. (1998). Guide to the identification of common clear plastic films. *Society for the Preservation of Natural History Collections (SPNHC) Newsletter*, **3**, Fall 1998, 1–4.

Williamson, C. J. (1999). Identifying Plastics. In *Plastics collecting and conserving* (A. Quye and C. J. Williamson, eds) pp. 136–137. NMS Publishing Ltd.

Degradation of plastics

Summary

It is essential to understand the factors causing degradation prior to developing an approach to the conservation of plastics. Chapter 6 defines degradation as it relates to plastics in collections, discusses how degradation can be recognized and the main factors involved in physical, chemical and biological degradation. The relative importance of various degradation factors is dependant on the type of polymer, additives and processing conditions. In addition, the function of plastics before collection and the conditions in which they are displayed and stored after collection affect their rate of degradation. Four plastics have been identified as being more vulnerable to degradation than others in a museum, namely cellulose nitrate, cellulose acetate, plasticized PVC and polyurethane foam. The most frequently seen degradation pathways for those plastics will be described in detail.

Degradation of a plastic is any change which has adverse effects on its properties or function. Industrial polymer chemists regard only the chemical and physical changes to a polymer to be important (McNeil, 1992). Degradation of plastics in museums is not defined by physical and chemical changes alone but by the resulting loss in function, form or significance of the object. Other terms used to describe degradation include 'ageing', which is usually associated with long-term changes due to weathering, and 'corrosion' borrowed from metal chemistry.

Degradation may occur during two phases in the life cycle of plastics. Firstly, during manufacture plastics are subjected to high temperatures under moulding and extruding, which provide opportunities for thermal and oxidative degradation. Secondly, during use plastic is exposed continually to air, moisture, light and heat (known as environmental weathering by polymer chemists) so changes in the chemistry of the polymer chains and additives are likely (Pagliarino, 1999). Most objects have been used or displayed prior to collection by museums. Their history contributes to their rate of degradation. There is rarely any information about the environments in which a museum object has been used or stored prior to collection. Prolonged exposure to light, heat, moisture, chemicals and gaseous pollutants during that period will reduce longevity.

The first publication concerning degradation of a commercial polymer appeared in 1861 in the *Journal of the Chemical Society* (Grassie, 1966). It concerned the failure of gutta percha cable insulation used to construct the East Indian telegraph cables, which deteriorated immediately after installation resulting in substantial financial loss. Gutta percha is an inelastic natural polymer produced by the tree *Palaquium gutta*. However, the plastics industry did not research degradation of synthetic polymers in a structured way until the 1960s and the subject is still at an early stage of development.

Condition surveys of plastics-containing objects in the V&A Museum (Then and Oakley, 1993) and British Museum (Shashoua and Ward, 1995) in the early 1990s indicated that around 1 per cent of objects were actively deteriorating and were unstable while 12 per cent showed degradation (Table 6.1). Most plastics examined were stable, displaying surface damage such as scratches and dirt. All the unstable objects contained cellulose nitrate, cellulose acetate, plasticized poly (vinyl chloride) (PVC) or polyurethane foam. Instability of the earliest plastics, cellulose nitrate and acetate, is expected due to their poorly stabilized and largely experimental formulations and because they are the oldest man-made plastics in museums. However, PVC and polyurethanes were first fully developed after World War II and are still in commercial use, so their degradation is less readily accepted.

Table 6.1 Condition of plastics in museums		
Plastics objects in defined condition (% of total surveyed)		
Condition	**V&A Museum (4500 plastics objects)**	**British Museum (3032 plastics objects)**
good and stable	>50	27.5
slightly damaged and stable	24	60
damaged and unstable	13	12
actively deteriorating and unstable	1	0.6

In general, degradation of plastics in museums is detectable by appearance, odour or feel, within 5–25 years of collection. The definition of useful lifetime as applied to plastics in museums is rather different to that used by the plastics industry. Quackenbos, an industrial chemist, defined the life of a plasticized

PVC film as the time taken for it to lose 10 per cent of its original weight (Quackenbos, 1954). After such loss, chemical properties of the material were considered to have changed so much that it failed. It has been argued that the useful lifetime for museum objects is reached when they cease to have a recognizable form or meaning (Bradley, 1994). In addition to establishing the degradation pathways for a plastic, it is also necessary to decide how much degradation is acceptable before the object loses quality. This is often more complex for modern art than for ethnographic materials. In addition, while yellowing and other changes in appearance are normal manifestations of degradation for natural materials and are usually left untreated, the same changes in plastics are usually deemed unacceptable (van Oosten, 1999).

The major causes of degradation of plastics may be attributed to physical, chemical and biological factors. Physical factors may arise from use prior to collection by museums, from interaction with the storage or display microclimate, and from migration of the additives incorporated during manufacture. Chemical causes of degradation comprise reaction of plastics with oxygen, ozone, water, metals, light and heat. Biological causes of degradation include micro-organisms and fungi. It is usual for a blend of physical and chemical factors to cause damage to plastics in museums, while biological degradation is less common.

This chapter will first describe types of degradation resulting from attack by various physical, chemical and biological factors. As mentioned previously, four plastics have been identified as being more vulnerable to degradation than others in museums: cellulose nitrate, cellulose acetate, plasticized PVC and polyurethane foam. The most frequently seen degradation pathways for those plastics will be described in detail. The final section in this chapter defines and illustrates (with examples) the definitions used by many museum professionals in Europe to describe degradation of plastics when completing a condition report.

6.1 Degradation attributed to physical factors

Physical factors include damage due to use prior to collection, changes in appearance and mechanical properties due to interaction with storage climate, and migration of the additives incorporated during manufacture.

6.1.1 Degradation due to mechanical use of plastics

Stress, fatigue and mechanical damage result from use of plastics usually prior to collection, although incorrect handling of museum objects can also be a factor. Such damage reduces the significance and function of objects (Figure 6.1). Repeated bending of a PVC soft toy leading to its failure is an example of physical degradation. Another is scratching a vinyl record by pushing a record player's needle across the grooves or fingering its surfaces, causing a distortion of sound when played. Scratches reduce the amount of light reflected from

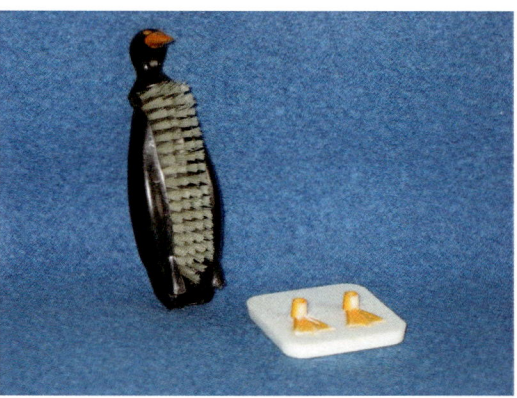

Figure 6.1

This polystyrene nailbrush in the form of a penguin was dropped, separating the supporting base from the body. The nailbrush dates from the 1960s, is 12 cm high, and has become brittle through exposure to light. The increased brittleness, due to chemical degradation, is likely to have contributed to its physical degradation. The brush cannot function as new and has lost its original form.

surfaces, thereby imparting a matt appearance. They also create microclimates by trapping moisture, particulates and pollutants which have the potential to cause chemical degradation locally.

6.1.2   Degradation due to physical interaction with surroundings

Although plastics are generally believed to be impermeable to gases, vapours and liquids, this is not true of all types. Polyethylene and polypropylene are effective at absorbing oily liquids, resulting in discolouration and changes to the feel and appearance of surfaces. Polyethylene Tupperware® food containers often develop tacky internal surfaces after long-term use due to the absorption of oily materials from foods. The tackiness introduced in the containers gives them a similar feel to PVC from which plasticizer has migrated, leading to frequent confusion in identifying the plastic. One cause of discolouration in polyethylene is absorption of coloured materials in close contact. Low density polyethylene manikin heads have been used to display gas masks since the 1950s by the Danish Royal Arsenal Museum in Copenhagen. The masks were fastened around the manikins to demonstrate usage. After around 20 years, the sulphur-containing vulcanizing (crosslinking) agent in the rubber straps of the gas masks had diffused into the polyethylene, irreversibly staining brown areas around the eyes, nose, chin and back of the head (Figures 6.2 and 6.3).

Change in temperature causes physical degradation of plastics. Heating plastics raises the kinetic energy of the polymer molecules, but their mechanical

Figure 6.2

Front view of polyethylene manikins from the 1950s used by the Danish Royal Arsenal Museum to display gas masks. Polyethylene has absorbed the vulcanizing agent from the gas masks' rubber straps. The vulcanizing agent contains sulphur and has stained the polyethylene irreversibly.

properties remain unchanged until the glass transition temperature (Tg) is reached. At the Tg, which is specific to each polymer type, a change from glassy to rubbery behaviour takes place, shown by dramatically increased flexibility. Below their Tg, polymers are hard and brittle, due to a lack of molecular mobility. Polyethylene, polypropylene and plasticized PVC have Tg values below ambient and are usually seen in their rubbery state in museums. Poly (methyl methacrylate) (Tg 45–115°C), polystyrene (Tg 80–104°C) and bisphenol A-based polycarbonates (Tg ca.150°C) are in a glassy state at room temperature (Brydson, 1999). If heated, they will become more flexible, flow and distort on handling. Vinyl records readily undergo physical degradation due to heating. The polymer from which vinyl records are manufactured

Figure 6.3
Side view of polyethylene manikin from the Danish Royal Arsenal Museum showing stains caused by vulcanizing agent in rubber.

(vinyl acetate/vinyl chloride copolymer) has a Tg of 30°C. At ambient temperature, the record is in its glassy state and holds its shape on handling. When warmed, for example by a nearby radiator, it readily distorts to assume a corrugated form (Figure 6.4). In theory the record may be returned to its original shape by reheating and reshaping. In practice it cannot be made to sound exactly as it did before heating because the grooves are no longer perfectly aligned.

On cooling, plastics tend to contract or shrink more than other materials found in museum collections including metals, ceramics and glass, and to

Figure 6.4

Single vinyl record before (left) and after physical degradation due to heating to 50°C for 30 minutes (right).

become stiffer than they are at ambient. The linear coefficient of expansion for thermoplastics is 5–10 times greater than those of most metals. A copper pipe will shrink by 0.01 per cent if the temperature is reduced by 10°C. Under the same conditions, a high density polyethylene pipe shrinks by 0.07 per cent, and polypropylene and unplasticized PVC pipes by 0.04 per cent (Nason et al., 1951). Although shrinkage of plastics is an unavoidable process on cooling, it is reversible. In the absence of degradation, the plastic assumes its original dimensions on return to ambient conditions.

However, cooling may be a cause of physical degradation in composites, that is objects which are constructed from several different materials in close contact. When composites are cooled, each material will attempt to shrink freely, but is restrained by its neighbours. Differential shrinkage on cooling may cause failure of one material unless the stress introduced by the temperature change is absorbed by elastic or inelastic deformation. The dimensional stability of photographic films, which are prepared from laminates of gelatine emulsion (containing information about the image) and plastic support layers, such as cellulose nitrate or acetate, has been extensively studied (Mecklenburg, 1994). The thermal coefficients of expansion of the emulsion and support layers are different, around $3.0 \times 10^{-5}°C^{-1}$ for photographic gelatine and $8.0–6.0 \times 10^{-5}°C^{-1}$ for cellulose acetate (Weast, 1974). Photographic film may undergo physical damage, including delamination and cracking on cooling, due to differential shrinkage of the various components (Shashoua, 2004) (Figure 6.5).

Stiffening is another physical change induced in plastics on cooling. Cellulose nitrate-based photographic film becomes measurably stiffer around −20°C, the temperature of domestic freezers, and breaks readily despite careful

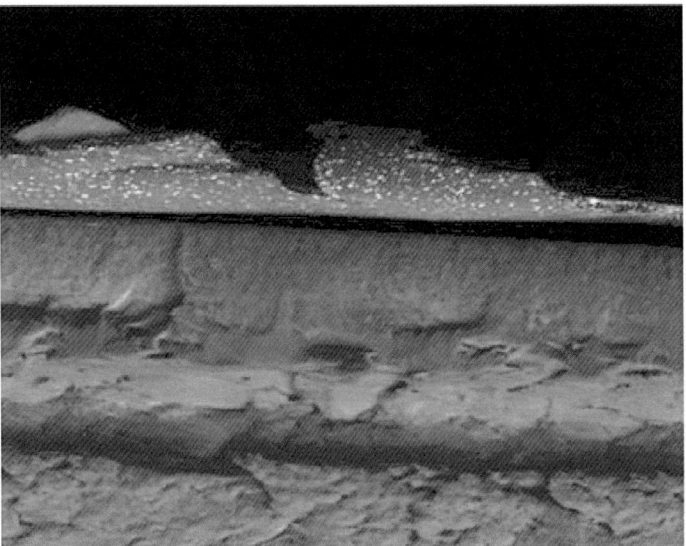

Figure 6.5
Low vacuum scanning electron micrograph of cellulose acetate-based photographic film cooled rapidly from 25°C to −5°C. Thin gelatine layer (upper) containing the silver particles (image) has separated from cellulose acetate support (lower) during cooling. Film is 0.5 mm thick.

handling while cold. The film regains its original flexibility on return to ambient. The flexibilities of the two major components of the film – cellulose nitrate and gelatine – are directly related to temperature. A sheet of cellulose nitrate in good condition shows an increase in Young's modulus (decrease in flexibility) of 21 per cent from 1378 MPa to 1654 MPa on cooling from 25°C to −5°C (Nason et al., 1951).

Water acts as a plasticizer for many of the early plastics. A sheet of casein, 4 mm thick, absorbs 5–7 per cent of its own weight in 24 hours and 30 per cent in 28 days under normal room conditions (Brydson, 1999). Polyamides, such as nylon, are the most hygroscopic polymers in common use today, containing up to 3 per cent moisture by weight under ambient conditions. In the same environment, cellulose acetate contains 0.8 per cent and poly (methyl methacrylate) and polystyrene 0.1 per cent. Plasticized PVC swells and appears opaque if stored at high relative humidity (RH) (Figure 6.6). Water vapour is an efficient plasticizer for the PVC polymer but is incompatible with commercial ester plasticizers, which are hydrophobic. As a result, the more plasticized the PVC, the less water it absorbs. Because water is only weakly bonded to the PVC polymer, it rapidly evaporates again.

Figure 6.6

Plasticized PVC tubing after storage at 30 per cent RH (left) and 90 per cent RH (right) for 24 hours. The opacity introduced at high relative humidity was reversed after 24 hours at 30 per cent. Tube dimensions 1.5 mm × 30 mm.

In addition to polymeric degradation, additives can be physically changed by moisture. Wood particles, paper fibres and flock were used as fillers to add reinforcement and bulk to formaldehyde plastics, notably Bakelite. Such fillers swell in contact with moisture thus causing the polymer to crack and fail.

6.1.3 Degradation due to migration of additives

The quantity of additives present in plastics formulations is related to the function, expected lifetime and price of a plastic. Additives may be exhausted or may migrate or evaporate from a plastic when its intended commercial lifetime is reached. Camphor, one of the earliest commercial plasticizers used until the 1930s, changes phase from solid to gas without forming a liquid (sublimes) at ambient temperatures and so is lost rapidly from cellulose nitrate formulations. Plasticizer loss causes shrinkage and brittleness. By contrast, liquid plasticizers which have relatively high boiling points may form sticky films at the surfaces of plastics prior to evaporating. The sticky surfaces trap dust, which may contain moisture and pollutants, resulting in subsequent chemical degradation of the polymer. Di (2-ethylhexy) phthalate (DEHP) has been the most frequently used plasticizer for PVC since the 1950s. DEHP often separates from the associated PVC polymer chains and is seen as a sticky, oily layer at surfaces prior to evaporating. DEHP has a boiling point of 386°C. Dust particles adhere readily to the sticky plasticizer film, obscure the original surfaces and change their appearances (Figures 6.7 and 6.8).

Figure 6.7

BASF's advertising model made from plasticized PVC in the 1960s has become sticky with time due to phthalate plasticizer at the surfaces. Dust from a nearby radiator has adhered to the hands and face.

Figure 6.8

Closer inspection of the face reveals that the dust adhered to the tacky PVC face both disfigures and changes the appearance of the model.

Most plasticized PVC formulations contain between 1–3 per cent by weight of a lubricant to prevent excessive adhesion to the mould during production. Adhesion slows production and results in a product with a faulty surface. The most common lubricant is stearic acid (Sears and Darby, 1982). Stearic acid is almost completely incompatible with PVC, due to its molecular structure. It is extremely non-polar, while PVC is polar due to the carbon-chlorine dipole. Degradation due to separation of stearic acid from PVC followed by migration of the lubricant to surfaces was exhibited by a hollow doll manufactured in the 1960s. After use, the doll had been stored in a cellar for 20 years in a cardboard box. On removal from storage, there was a dramatic change in the doll's appearance. A dense, white, brittle layer identified as stearic acid obscured 90 per cent of the skin area on both legs. The body under the dress, neck, face, arms and hands showed only discrete patches of the white material (Figure 6.9).

Figure 6.9
Doll exhibiting white crystalline stearic acid lubricant on legs, body and arms. The doll is 40 cm high.

6.2   Degradation attributed to chemical factors

Chemical causes of degradation comprise reaction of plastics with oxygen, ozone, water, metals and radiation mainly as visible light or heat. These factors provide sufficient energy and appropriate environments to break selected chemical bonds present in polymers and additives. In general, degradation results in the reduction of molecular weight and the formation of new chemical structures. Conditions which accelerate degradation may be present during manufacture, use, display or storage. The extent and type of reaction of plastics to environmental factors depends on their chemical structure (Table 6.2).

Table 6.2 Effect of radiation, heat, oxygen and water on the major plastics in collections					
Plastic		**Climatic factors**			
Type	**Example**	**Ultraviolet radiation and light**	**Heat**	**Oxygen**	**Water**
semi-synthetics *cellulose esters*	cellulose nitrate	discolours, crazes	softens on warming; Tg 70°C; loss of plasticizer; camphor sublimes at 20°C	photo-oxidation	acid and alkaline hydrolysis; nitrous and sulphuric acids produced
	cellulose acetates	discolours	softens on warming; Tg 50°C; melting point 96°C	photo-oxidation	acid hydrolysis; acetic acid produced
protein derived	casein-formaldehyde	Discolours	softens on warming to 80–85°C	photo-oxidation causes crazing and discolouration	water softens and swells
synthetics *phenolics*	phenol-formaldehyde	reduction in surface lustre	thermosetting; degrades above 260°C	thermal and photo-oxidation causes surfaces to powder	polymer resistant to hydrolysis; cellulose filler swells
aminoplastics	amino-, thiourea-, urea- and melamine-formaldehydes	reduction in surface lustre	thermosetting, distorts around 140°C	thermal- and photo-oxidation	polymer crazes on prolonged contact
vinyls	poly (vinyl chloride) (rigid)	discolours from yellow to orange, red, brown and black due with loss of hydrogen chloride	softens on warming; Tg 80°C	thermal- and photo-oxidation cause loss of hydrogen chloride and discolouration	water softens and opacifies PVC polymer reversibly

Table 6.2 Continued					
Plastic		**Climate factors**			
Type	**Example**	**Ultraviolet radiation and light**	**Heat**	**Oxygen**	**Water**
	poly (vinyl chloride) (plasticized)	discolouration of PVC polymer	softens on warming; Tg −25 to +25°C, depending on percentage plasticizer; plasticizer diffuses to surfaces and evaporates; PVC discolours	thermal- and photo-oxidation of phthalate plasticizers in acid to phthalic acid crystals	phthalate plasticizers hydrolyzed by acid to phthalic acid crystals
polyolefins	polyethylene (low density polyethylene)	yellows readily; becomes brittle	softens on warming; Tg −20°C; melting temperature 109–120°C	photo-oxidation causes discolouration; thermal oxidation results in crosslinking and stiffening above 50°C	deformed and sometimes stress cracked by boiling water
	polypropylene	discolours and becomes brittle	softens on warming; Tg 5°C; melting temperature 150–170°C	photo-oxidation causes discolouration; thermal oxidation results in chain scission and disintegration	resistant to moisture
polyamides	nylon 6	yellows and becomes brittle	softens on warming thermoplastic; Tg ca.75°C; distorts above 60°C under pressure	nylon 6 resistant to oxidation; other nylons photo-oxidized and become brittle	hydrolyzed by alkaline conditions
acrylics	poly (methylmethacrylate)	yellowing and opacity of transparent materials	softens on warming; Tg 50°C; melts at 100°C	photo-oxidation causes discolouration	resistant to moisture
polystyrene	polystyrene	yellows and becomes brittle	softens on warming; Tg 100°C; melting temperature 230°C	photo-oxidation increases stiffening by crosslinking	resistant to moisture

Plastic		Climatic factors			
Type	Example	Ultraviolet radiation and light	Heat	Oxygen	Water
	acrylonitrile-butadiene-styrene (ABS)	yellows due to butadiene component and becomes brittle	softens on warming; Tg 100°C; depolymerizes above 350°C	photo-oxidation causes stiffening	resistant to moisture
polyesters	poly (ethylene terephthalate)	yellows	softens on warming; Tg 50–70°C; melting temperature 265°C	photo-oxidation causes discolouration	hydrolyzed by alkaline conditions
polyurethanes	polyurethane-polyester-based (PUR ester)	discolours and becomes brittle	thermoplastic and thermosetting types; loss of properties above c.80°C	resistant to oxidation	hydrolysis results in chain scission and crumbling
	polyurethane-polyether-based (PUR ether)	discolours and darkens; becomes brittle	thermoplastic and thermosetting types; loss of properties above c.100°C	photo-oxidation causes brittleness; thermal oxidation produces isocyanates and polyols	resistant to moisture
epoxies (epoxide)	epichlorohydrin/bisphenol-A	yellows and becomes brittle	thermosetting; loss in properties above 175°C	photo-oxidation causes loss of gloss and opacity	hydrolysis of polyester-based epoxies
polycarbonates	polycarbonate of bisphenol-A	yellows readily due to impurities in bisphenol-A	softens on warming; Tg 150°C; melts at 225°C	resistant to oxidation under ambient conditions	resistant to moisture

Table 6.2   Continued

Changes in chemical structure caused by degradation, particularly changes to molecular weight, usually increase plastics' sensitivities to moisture and temperature.

The structural changes introduced into polymers as a result of chemical degradation may be divided into chain scission, crosslinking, development of chromophores and development of polar groups. The term 'chromophore' is derived from the Greek word for 'colour bringer'. A chromophore is a molecule that imparts colour by selectively absorbing light at particular wavelengths.

Chain scission

Chain scission is the breaking of a polymer chain to create a polymer with two or more shorter chains. It is the most common structural change caused by chemical degradation. Because many physical and chemical properties are dependent on molecular weight, chain scission often results in a catastrophic loss in strength and toughness. The molecular weight of the polymeric component of plastics is very high, ranging from around 30 thousand for nylons and polyesters to more than one million for polyethylene. The breaking of just one out of every 10 000 bonds in a polymer backbone will dramatically reduce its molecular weight and associated physical properties. Failure of the plastic will result if the degradation reaction continues unhindered. An example of the manifestation of chain scission is the crumbling of polyurethane foams.

Crosslinking

The formation of crosslinks in a linear polymer causes two polymer chains to join, resulting in an increase in molecular weight. Linear polymers including polystyrene and PVC are most susceptible to crosslinking. Such degradation is detected as increased stiffness, brittleness and a decrease in solubility as the average molecular weight increases.

Development of chromophoric (light-absorbing) groups

The formation of chromophoric groups such as carbonyls ($C{=}O$) and unsaturated carbon bonds ($C{=}C$ and $C{\equiv}C$) in the backbone of a polymer results in colour formation in white and transparent plastics and discolouration of coloured materials. Chromophoric groups can be formed when polymers undergo oxidation (see section 6.2.3) or hydrolysis (see section 6.2.4).

Development of polar groups

A polar group is a chemical structure in which the distribution of electrons is uneven, enabling it to take part in electrostatic interactions. Polar groups in polymers include hydroxyl (O—H), carbon-hydrogen (C—H) and carbon-chlorine (C—Cl). The formation of polar groups in the polymer backbone or side chains due to oxidation, hydrolysis and other reactions associated with degradation changes the chemical reactivity of polymers and may affect some physical properties including solubility. It is not only the chemical structures of plastics which control the rate and extent of chemical degradation, but also their physical properties. Oxygen and water diffuse more readily through amorphous (unstructured polymer chains) than through crystalline (organized polymer chains) materials, resulting in more rapid oxidation and hydrolysis in the former. Polymers with Tg below ambient are in a rubbery state, exhibit high molecular mobility and allow movement of reactive species and degradation products through the material more readily than polymers with higher Tg values. The greater mobility of species involved in degradation also increases the rates of reaction.

6.2.1   Degradation due to chemical reaction with light

Light, particularly ultraviolet (UV) light with wavelengths 200–800 nm, is considered the factor most damaging to plastics. Energy, **E**, provided by a photon with a particular wavelength, can be calculated from the equation:

$$\mathbf{E} = \frac{hc}{\lambda}$$

where h is Planck's constant, 6.63×10^{-34} J s
c is the velocity of light in a vacuum 3.00×10^8 ms^{-1}
λ is the wavelength of light.

The shorter the wavelength of radiation, the more energy it contains (Figure 6.10). Because bond energies in polymer molecules are typically between 300–500 kJ per mole, they are resistant to visible and infrared radiation, which exhibit maximum energies 300 and 180 kJ mol^{-1} respectively. UV light with wavelengths lower than 400 nm contains sufficient energy to break C—C, C—O and C—Cl bonds but not C—H or C—F (McNeill, 1992).

Figure 6.10

Energy provided by ultraviolet, visible and near infrared radiation.

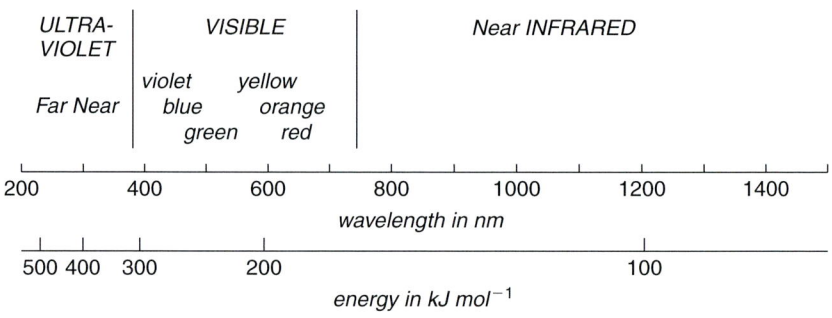

In addition to introducing chemical changes in polymers, light also causes dyes and colorants to fade (Figure 6.11).

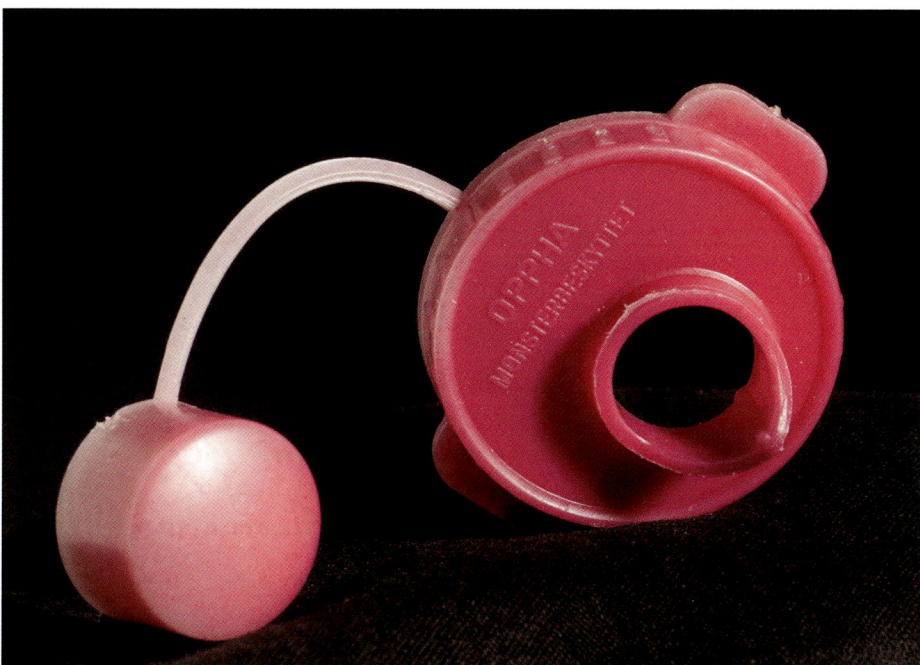

Figure 6.11
Dye used to colour a
polyethylene Danish
milk bottle cap
from the 1960s has
faded unevenly after
prolonged exposure to
sunlight.

Box 6.1   Chemistry of reactions between light and polymers

The primary pathway by which light can react with polymers is for it to be absorbed
by particular functional groups known as chromophores (light-absorbing groups).
Chromophoric groups may either be chemical structures in the polymer backbone
or may be present as impurities from manufacture. Unsaturated groups including
$C{=}O$ (carbonyl) and $C{=}C$ have absorption maxima between 200 and 400 nm and
are those most susceptible to photodegradation. Carbonyl groups in polyesters,
nylons, polycarbonates and polyurethanes absorb specific wavelengths in the UV
spectrum so weakly that any remaining light energy continues into the bulk of the
plastic before its energy is used up. It is thought that absorption of light results in
the formation of free radicals via mechanisms known as Norrish Type I.

$$—CH_2CH_2\mathbf{C}CH_2CH_2CH_2— \longrightarrow —CH_2CH_2C^\bullet + {}^\bullet CH_2CH_2CH_2—$$

$$\underset{\mathbf{O}}{\|} \qquad\qquad\qquad \underset{O}{\|}$$

(**Bold** = Free radical site)

Surprisingly, some polymers which have no chromophores in their structure
such as PVC, polyethylene, poly (methyl methacrylate), nylon and polystyrene are
degraded by exposure to radiation with wavelengths around 200 nm, undergoing
discolouration and weakening. This may be attributed to the formation of
chromophoric carbonyl groups, by oxidation. Impurities or additives in plastics may
also contribute chromophores.

An alternative pathway by which light can react with polymers is via compounds known as sensitizers, which absorb radiation, become energetically excited and transfer the absorbed energy to other molecules. This pathway has poorly studied. Sensitizers include aromatic ketones such as acetophenone ($C_6H_5COCH_3$) and benzophenone ($C_6H_5COC_6H_5$). Sensitizers can also transfer energy to oxygen to produce reactive singlet oxygen, which reacts further. Weak links, particularly at chain ends, can be the site for 'unzipping' reactions. A monomer or other simple molecule is removed leaving the new chain end unstable. This process continues and results in depolymerization. Yellowing and discolouration are common manifestations of this form of photodegradation and are attributed to the formation of conjugated structures, including polyenes (alternate carbon-carbon double bonds and carbon-carbon single bonds) or carbon-carbon double bonds conjugated to carbonyl or aromatic groups.

6.2.2 Degradation due to chemical reaction with heat (thermolysis)

Thermal degradation, also known as thermolysis, is a process by which the action of heat results in reduction of physical, chemical or electrical properties. Thermal degradation should be distinguished from thermal decomposition, a process involving the formation of flammable vapours and extensive changes in chemical species caused by fire or high temperatures. Thermal decomposition will not be discussed here since it is a rare occurrence in museums and is covered in detail by other publications (Beyler and Hirschler, 2002). Heat is the most likely cause of degradation for plastics in museum collections since light is largely excluded in storage areas and is filtered of high energy ultraviolet radiation in exhibitions. All polymers undergo degradation if exposed to sufficiently high temperatures, but the range of tolerated temperatures is wide. Unstabilized, unplasticized PVC discolours around 70°C whereas polytetrafluoroethylene can resist temperatures up to 500°C.

A long-standing rule of thumb stating that increasing the temperature by 10°C doubles the rate of thermal reactions has recently been developed further to relate better to unstable materials in museums. Michalski (2002) suggests that a 5°C increase in temperature doubles the rate of thermal degradation of polymers in museums, thereby halving their chemical lifetimes.

Heating plastics provides sufficient energy to break bonds in the backbones or side chains of polymers in the same way as light. Depolymerization occurs when the backbone of a polymer breaks, first forming small molecules such as dimers (two monomer units) and trimers (three monomer units) or other fragments with shorter chains than the original polymer. These break down further to form monomers via a process known as depropagation. Poly (methyl methacrylate) undergoes depolymerization followed by depropagation to pure monomer. In addition to monomer, polystyrene also yields dimer, trimer and chain fragments on heating. Because many physical and chemical properties are related to the

Box 6.2   Chemistry of reactions between heat and polymers

Predicting which bonds, if any, are likely to be broken on heating is not a simple process. Not all sites in a polymer are equally vulnerable to thermolysis. The weakest bonds in the polymer are those most likely to break. Bonds in impurities which have become incorporated into polymer molecules, including traces of monomers and residues of catalysts from synthesis, are more vulnerable than the same bonds in the polymer's backbone. Terminal groups and O—O bonds formed as a product of oxidation are also more vulnerable than the same bonds located in the backbone.

The dissociation energy of bonds present in a polymer provides a guide to its thermal stability (Table 6.3). The fraction of bonds with sufficient energy to dissociate can be calculated from the Boltzmann factor:

Table 6.3   Mean dissociation energies of saturated and unsaturated bonds (kJ mol^{-1})

A-B	H	C	N	O	F	Cl	Br	I	S	Si
H		416	391	464	563	432	366	299	339	
C	416	348	292	361	485	329	280	240	259	328
N	391	307	160	222	270	200				
O	464	361		139	185	203				369
F	563	485			153	254				
Cl		340				243			250	
Br		280					219	193	178	
I								151		
S		273							213	
Si	319	328		432	541	359				226
A=B										
C		612		732						
N		617	418							
O		732	607	489						
A≡B										
C		812								
N		891	946							

Fraction of bonds with dissociation energy = exp (-D/RT)

where:

D = dissociation energy
T = temperature in degrees Kelvin
R = universal gas constant

Using this relation, it can be calculated that the temperature required for one mole of C—C bonds to dissociate into radicals is 486°C, while one mole of O—O bonds requires only 30°C. Bond energies for unsaturated bonds are generally higher than their saturated equivalents. However, bond energies are not always reliable tools to predict the pathway followed by degradation. Although bond energies of C—Cl and C—C bonds are similar (340 and 348 kJ mol^{-1} respectively), PVC exhibits measurably lower thermal stability than polystyrene or polyethylene. The activation energy for thermal degradation of the PVC polymer is approximately 83.6 kJ/mol. This is low compared with that for polyethylene (192.5 kJ/mol), polystyrene (230.1 kJ/mol) and polypropylene (272 kJ/mol) (Rice and Adam, 1977). This difference may be attributed to structural abnormalities in PVC where deviation from the repeat structure occurs. Such deviations may occur at chain ends or may be introduced by processing and are often the sites where degradation starts.

average molecular weight of a polymer, depolymerization results in a measurable loss in mechanical properties including strength and toughness.

Reactions at side groups often take place at lower temperatures than those for backbone scission, so occur more readily. Side group scission involves the loss of low molecular weight fragments, which are mobile and can readily react with the original polymer and other materials in the vicinity. Heating causes scission in side groups and results in the loss of small molecules, which then destabilizes adjacent groups. This form of degradation is exhibited by PVC and poly (vinyl acetate). The loss of one molecule of acid (hydrogen chloride and acetic acid respectively) from each repeat unit destabilizes its neighbour, which also loses a molecule of acid. If the acid produced is not removed, the reaction becomes autocatalytic and accelerates (Brydson, 1999). The reaction continues along the chain to form alternate double bonds with single bonds, a structure known as a conjugated polyene (Figure 6.12).

As degradation proceeds, polymers begin to absorb radiation in the ultraviolet part of the spectrum and discolour partly because of the increasing number of conjugated polyenes but also due to the presence of coloured carbonium ions (carbon atoms with localized positive charges). After between 7–11 repeat units have formed, the polymer absorbs radiation at longer wavelengths until it is absorbing in the violet, blue and green parts of the spectrum. As polymers degrade, they change colour from white to yellow to orange to red, brown and, ultimately black.

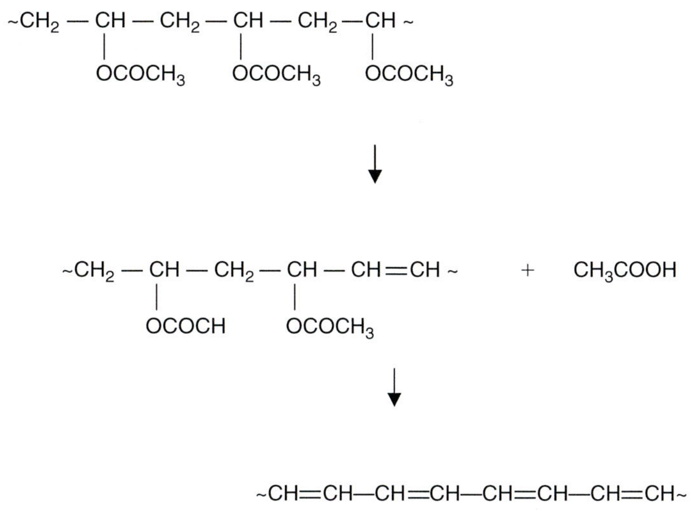

~CH$_2$ — CH — CH$_2$ — CH — CH$_2$ —CH ~
       | | |
     OCOCH$_3$ OCOCH$_3$ OCOCH$_3$

↓

~CH$_2$ — CH — CH$_2$ — CH — CH=CH ~   +   CH$_3$COOH
       | |
     OCOCH   OCOCH$_3$

↓

~CH=CH—CH=CH—CH=CH—CH=CH~

Figure 6.12
Side group scission of poly (vinyl acetate) produces a conjugated polyene and acetic acid.

6.2.3   *Degradation due to chemical reaction with oxygen*

Most degradation reaction paths for plastics involve oxygen. Polymers and additives can either react directly with molecular oxygen (known as autoxidation) or react with another reactive material derived from oxygen. Highly unstable and, therefore, highly reactive, ozone (O_3) is a product of the reaction between oxygen and ultraviolet light and causes oxidation. Oxygen comprises 21 per cent of the atmosphere, so contact between plastics and oxygen is inevitable. Polymers are subjected to oxidation at every stage in their life cycle – during manufacture and processing as well as during storage and end use. Plastics are frequently heated to between 150–250°C during processing and may be in a rubbery or melted state where they can readily react with oxygen. Metal particles from moulds or rollers can contaminate warm plastics and catalyze oxidation.

For chemical reaction with oxygen to take place, oxygen must first diffuse into solid plastics. Because of differences in their oxygen permeabilities, crystalline polymers such as polyethylene and polytetrafluoroethylene are more resistant to oxidation than amorphous materials such as polystyrene and poly (methyl methacrylate). Not all chemical bonds in a polymer are equally reactive towards oxygen molecules. Unsaturated bonds (double or triple) react more readily with oxygen and ozone than saturated bonds (single). Branched chains containing a tertiary carbon atom (carbon atom attached to three other carbon atoms by single bonds) are reactive sites for oxidation. Tertiary carbons are slightly electronegative compared with the rest of the carbon chain and react readily with electropositive molecules such as oxygen.

Although oxidation of polymers can take place under ambient temperatures and in the absence of light, reactions are accelerated by the energy from both heat and light and the presence of metals, whether present as impurities or as part of the polymer chain. The source of energy and the microclimate affect the rate of oxidation for a particular plastic. Oxidation of polyethylene takes place at temperatures as low as 50°C, but can occur at room temperature if the polymer is exposed to ultraviolet light at the same time. The purer polyethylene, the slower it tends to oxidize in the presence of light because any chromophoric impurities absorb and react with light energy.

Once oxidation starts, it sets off a chain reaction which continues until no polymer remains. The chain reaction consists of three basic stages – initiation, chain propagation and chain termination. There is no single route by which each stage occurs, but the induction period is slow, an autocatalytic, fast oxidation stage follows and the rate slows again for the termination step (Sears and Darby, 1982). The first or initiation step involves generation of highly reactive free radicals ($\mathbf{R}^{\bullet}$). The free radicals are created when the two unpaired electrons in the ground state of the oxygen molecule remove the most reactive hydrogen atom from the polymer ($\mathbf{R}$).

$$polymer\ \mathbf{R} \xrightarrow{\substack{heat, light, \\ radiation}} 2\mathbf{R}^{\bullet}\ pair\ of\ free\ radicals$$

The free radicals subsequently react with oxygen molecules to produce hydroperoxy radicals ($\mathbf{RO}_2^{\bullet}$) and with the polymer itself to form hydroperoxides ($\mathbf{ROOH}$). This step is known as the propagation stage because the unstable free radicals formed enable degradation to continue:

$$\mathbf{R}^{\bullet} + O_2 \rightarrow RO_2^{\bullet}$$
$$RO_2^{\bullet} + RH \rightarrow ROOH + R^{\bullet}$$

The primary oxidation product, hydroperoxide $\mathbf{ROOH}$, decomposes by scission of the oxygen-oxygen bond to produce two radicals which can subsequently initiate reaction with oxygen.

$$ROOH \xrightarrow{heat, light} RO^{\bullet} + {}^{\bullet}OH$$
$$RO^{\bullet} \rightarrow R + ketones$$
$${}^{\bullet}OH + RH \rightarrow H_2O + R^{\bullet}$$

Oxidation can be stopped by a termination step if radicals react with each other to form non-radical species, especially esters and ketones which contain carbonyl groups.

$$\left.\begin{array}{l} RO_2 + RO_2 \\ RO_2 + R \\ R + R \end{array}\right\} \rightarrow \text{non-radical products}$$

Oxidation reduces the physical properties of plastics as well as changing their aesthetic appearance. Oxidation of polymers manifests as reduction in tensile strength, flexibility and fragmentation because of the resulting loss in molecular weight (Figure 6.13). Discolouration is often symptomatic of oxidation and results from the formation of chromophoric carbonyl groups from, for example, ketones. Oxidation can also result in disruption of surfaces shown by cracking, increased porosity and enhanced ability to absorb liquids compared with new material (Figure 6.14).

Oxidation does not proceed at the same rate throughout the lifetime of a polymer. The rates have been studied by quantifying either changes in the deteriorating polymer or changes in the microclimate with time. Real time degradation of plastics observed in museums is difficult to reproduce or model for study. For the convenience of research scientists, the useful lifetime of a polymer from manufacture to failure is frequently concentrated to a period of a few days, weeks or months by applying accelerated thermal, light or high relative humidity ageing or a combination of all three. However it is difficult to ascertain whether accelerated ageing represents oxidation in real time. Higher temperatures and light levels may initiate reactions not usually seen under ambient conditions.

Changes in colour, gloss, viscosity (directly related to molecular weight), physical properties (including tensile strength) and development or reduction of the concentration of particular chemical structures have all been used to follow the progress of oxidation (Williams, 1993). The amount of oxygen absorbed (oxygen uptake) from the surrounding atmosphere at a particular temperature by the polymer with time has been determined for museum objects containing polyethylene, PVC, poly (acrylonitrile-butadiene-styrene) and polyurethane (Grattan, 1993).

Changes in properties rarely occur at the same rate throughout a polymer's life cycle from manufacture to failure. This may be explained by the varying levels of importance of those chemical reactions causing degradation at different points in the polymer's history. Feller (1977) proposed that autoxidation occurs in four stages known as inception, induction, maximum rate/steady state and declining rate. During the inception stage, the polymer adsorbs oxygen rapidly and briefly. It has been attributed to the presence of weak links in the polymer, which break more rapidly than those in the backbone. The first section of the induction period is typified by few measurable changes in physical properties but, in practice, it represents the last opportunity to preserve the deteriorating polymer's original physical properties, so should be extended as long as possible. The second section of the induction stage is recognized by rapid, irreversible, autocatalytic degradation. The rate of oxidation remains high in the third stage of degradation and results in breakdown of many polymers due to brittleness, weakening and discolouration. During the final stage of oxidation, the rate slows probably because all the available reaction sites in the polymer have

Figure 6.13
Polyethylene carrier bag exhibiting typical signs of oxidation after 2 years in a dark storage area. The bag has fragmented and weakened due to a reduction in molecular weight.

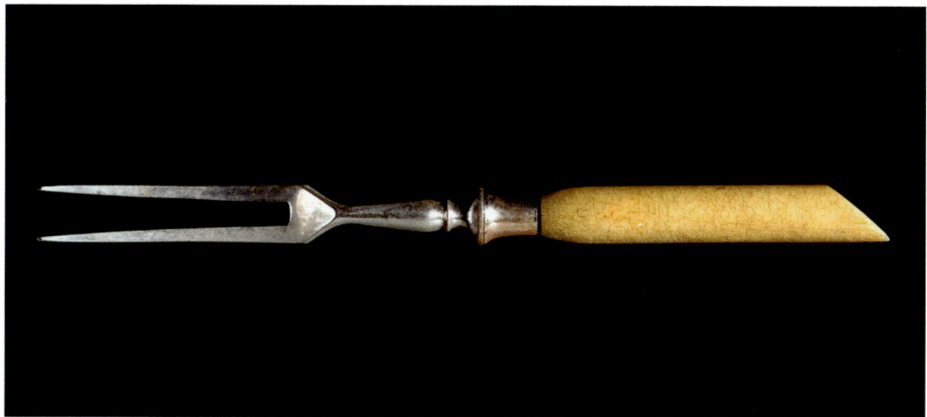

Figure 6.14
Oxidation in the presence of light has caused a casein-formaldehyde fork handle from
the 1940s to craze, crack, shrink and discolour.

been used. Increasing the temperature or intensity of light may cause further
reaction with oxygen.

6.2.4   Degradation due to chemical reaction with water

Some polymers are synthesized by joining monomers with a loss of water molecules, a process known as condensation polymerization. The reverse reaction is
achieved when bonds break with the addition of water molecules (hydrolysis or
hydrolytic scission). Hydrolysis is usually initiated by acidic or alkaline environments. Cellulosics such as cellulose nitrate and acetate, polyesters and polyester-
based polyurethanes are particularly vulnerable to hydrolysis. Because polymers tend to become more polar when oxidized, their sensitivity to water
increases on ageing. Chain scission occurs by the breaking of bonds at numerous sites along the polymeric backbone, resulting in a reduction in molecular
weight and loss of mechanical properties. Highly hydrophobic polymers such
as polyethylene and polypropylene are unlikely to have hydrolysable chemical
groups, so are not subject to hydrolytic breakdown.

6.2.5   Degradation due to chemical reaction with metals

Traces of metals are present in most polymers as a result of processing in stainless steel reaction vessels, being transported in metal containers and shaping
in metal moulds or rolling between metal rollers. Many metal ions catalyze oxidation by accelerating decomposition of hydroperoxides at room temperature.
Observation of plastics materials in museums suggests that copper is a highly
effective catalyst for semi-synthetic materials such as cellulose nitrate and

acetate. The effectiveness of metals as catalysts depends on their oxidation states and follows the order (Rychly and Strlic, 2005):

$$Cu(II) > Cr(III) > Co(II) > Fe(III) > Mn(II) > Ni(II)$$

The influence of other metals on the rate of degradation is unclear. Some authors suggest that aluminium, titanium, zinc and vanadium ions act to slow the rate of oxidation. Others suggest that while titanium white (anatase form) and zinc oxide in acrylic paints initially slow crosslinking by absorbing ultraviolet radiation, the metal compounds subsequently accelerate oxidation in the presence of light and water leading to the formation of hydrogen peroxide (H_2O_2). This, in turn, forms hydroxyl radicals (Mills and White, 1994):

$$ZnO + O_2 \rightarrow (ZnO)^+ + O_2^\bullet -$$
$$O_2^\bullet + H_2O \rightarrow HO_2^\bullet + HO^-$$
$$2HO_2^\bullet \rightarrow H_2O_2 + O_2$$
$$H_2O_2 \rightarrow 2HO^\bullet$$

6.3   Degradation attributed to biological factors

Conservation literature asserts that synthetic polymers are not themselves prone to biological degradation, although oil, protein and cellulose residues in contact with polymers may support bacteria and fungi (McNeill, 1992: Rivers and Umney, 2003). Biological degradation occurs when microbial enzymes chemically break down large molecules into smaller fragments which are water-soluble. Wiles suggests that no fungus has yet developed enzymes to break down synthetic polymers since they don't exist in nature (Wiles, 1993). In contrast, other researchers propose that selected synthetic polymers are subject to attack by micro-organisms. The reason for this discrepancy may be the various techniques used to evaluate the resistance of synthetic materials to micro-organisms, particularly the media used to support polymers during investigation and the addition or exclusion of glucose as a source of nutrition for bacteria.

Acrylic polymers, including poly (ethyl methacrylate), are subject to biodegradation, although many commercially available acrylic paints and coatings are supplied with biocide in the formulation so do not support attack (Cappitelli et al., 2004). Examination by optical and scanning electron microscopies suggests that fungi grow readily on poly (vinyl acetate) and less easily on acrylic polymers. Fungal attack has also been reported on deteriorated cellulose nitrate in the form of a fake leather surface of a carrying case for a film projector dating from the 1960s (Silva, 2006). Infrared spectroscopy and optical microscopy were used to identify fungal residues. Because cellulose nitrate forms nitric acid during degradation, it is not expected to sustain biological degradation. Micro-organisms attack plasticizers in soft PVC rather than the PVC polymer because carbon is more readily available in the lower molecular weight plasticizer (Moriyama et al., 1993). Cappitelli suggests that biomass grown on plasticizers produces enzymes which subsequently attack the polymer component (Cappitelli et al., 2004).

6.4 Degradation pathways for the least stable plastics in museum collections

The four plastic types known to deteriorate most rapidly in museum collections are cellulose nitrate, cellulose acetate, plasticized PVC and polyurethane foams. During degradation, these plastics produce gaseous products which initiate or accelerate degradation of other organic materials and corrode metals in the vicinity. They have been described as 'malignant plastics' (Williams, 2002). Williams calls plastics which degrade without offgassing corrosive products 'benign'.

6.4.1 Degradation of cellulose nitrate

Cellulose nitrate (CN) is broken down by both physical and chemical factors. Loss of plasticizer causes shrinkage and cracking. Until the 1930s, the plasticizer used to soften CN for commercial use was camphor (1,7,7-trimethyl-bicyclo [2,2,1] hepta-2-one), which sublimes (changes from solid to vapour without forming a liquid) at room temperature. The vapour pressure of camphor is sufficiently high at around 0.013 Pa at its melting point (179°C) that it can sublime under ambient conditions. Mixtures of camphor and CN in a 1:2 weight ratio, known as celluloid, lose plasticizer slowly and contain around 15 per cent by weight after 30–40 years (Selwitz, 1988). Shortly before World War II, camphor was replaced by triphenyl and tricresyl phosphates and phthalates. Dibutyl phthalate and di (2-ethylhexyl phthalate) have been used since the 1920s. It is likely that plasticizers also act as stabilizers, perhaps by adsorbing the acidic degradation products evolved by the CN polymer. When plasticizer evaporates, the CN polymer becomes more vulnerable to chemical degradation (Shashoua et al., 1992) (Figure 6.15). At this early stage of degradation, CN is highly flammable and burns at temperatures up to 15 times higher than those attained by burning paper.

The CN polymer undergoes thermal, photochemical and hydrolytic degradation reactions, the latter being the most important (Hamrang, 1994). In addition, breakdown of the polymer is autocatalytic. If not removed from the undegraded material, the breakdown products catalyze a faster and more

Figure 6.15

An early stage of degradation of cellulose nitrate. A poster from the Historical Music Museum in Copenhagen, Denmark, dating from the 1950s has lost camphor and shows cracks. At this stage, the poster is highly flammable.

extensive reaction than the primary processes. This may happen if a pair of spectacles with cellulose nitrate frames were stored enclosed in their case, for example. The major product of thermal degradation is the highly reactive oxidizing agent nitrogen dioxide (NO_2) identified by its yellow vapour and distinctive odour. Nitrogen dioxide is formed by cleavage of the N—O bonds in the secondary nitrate group joining the cellulose ring at positions 2 and 3, which at $167\,kJmol^{-1}$ are the weakest bonds in the molecule. Equivalent bonds at the primary position 6 have a bond energy of $330\,kJmol^{-1}$ which is similar to bond energies in the backbone (Figure 6.16). Relating increasing concentration of carbonyl groups formed as a result of the cleavage of the N—O bonds with the decreasing concentration of N—O during thermal degradation has been studied using FTIR spectroscopy (Shashoua et al., 1992).

Figure 6.16

Thermal breakdown of cellulose nitrate. The figure shows a single cellulose ring or repeat unit, but many identical units comprise a cellulose molecule.

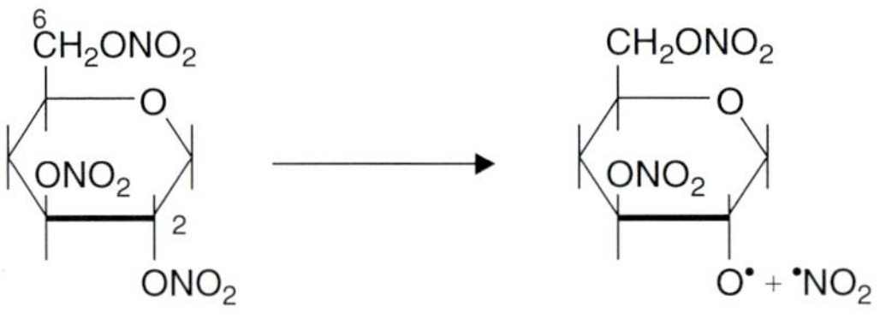

Nitrogen dioxide reacts with moisture in air to form nitric acid. Investigations using ion chromatography combined with infrared spectroscopy suggest that concentrations of sulphate ions higher than $5\,mg$ per gram cellulose nitrate, resulting from inadequate washing during manufacture, accelerate the production of NO_2 and the rate of discolouration of CN on thermal/high relative humidity ageing compared with samples containing no sulphate (Stewart et al.,

1996). The same study suggested that CN filled with zinc oxide deteriorated more slowly than unfilled material. It was proposed that the filler acts as a buffer, neutralizing acids and hence slowing degradation.

Water diffuses into cracks at surfaces formed when plasticizer evaporates, reacts with nitrogen dioxides evolved in the bulk of the CN object, and produces nitric acid there. The acid attacks the cellulose polymer chains resulting in chain scission along the backbone between the cellulose rings. A reduction in molecular weight follows, which is manifested typically by the formation of a network of cracks which start inside the object before spreading to surfaces causing brittleness and weakening. The formation of internal cracks before surface cracks in cellulose nitrate contrasts with the opposite pattern shown by synthetic plastics and often gives degraded CN objects the feel of intact material. As degradation continues, internal cracks or crazes develop and cellulose nitrate yellows (Figure 6.17). In the final stage, crazing known as crizzling is so extensive that cellulose nitrate disintegrates. At this point, its flammability is the same as that of paper.

Figure 6.17
A cellulose nitrate ruler displays extensive internal crazing, but can still be handled and used.

Cellulose nitrate is particularly susceptible to light of wavelengths 360–400 nm (Parkins, 1957). Degradation is due to a nitrate ester cleavage in a similar manner to thermal decomposition and is likely to result in the formation of a carbonyl group:

$$RCHONO_2 \rightarrow {}^\bullet NO_2 + RCHO^\bullet$$
$$RCHO^\bullet + RCHONO_2 \rightarrow RCHOH + RCONO_2^\bullet$$

$$^\bullet NO_2 + RCHONO_2 \rightarrow decomposition$$
$$RCONO_2^\bullet \rightarrow RC{=}O + {}^\bullet NO_2$$

At shorter wavelengths (those with higher energy) disintegration of the cellulose ring occurs, causing a rapid decrease in molecular weight. Once started this process continues even in the dark. Some metals, particularly copper, accelerate the rate of degradation of cellulose nitrate. Nitric acid produced by CN corrodes metals. If the metal is a structural component of the object, such as the shaft of a knife, corrosion can lead to destruction of the handle. With the corrosion layer, the volume occupied by the metal increases until the CN cracks and bursts (Figure 6.18). Nitric acid may also corrode metals in the vicinity of the CN object.

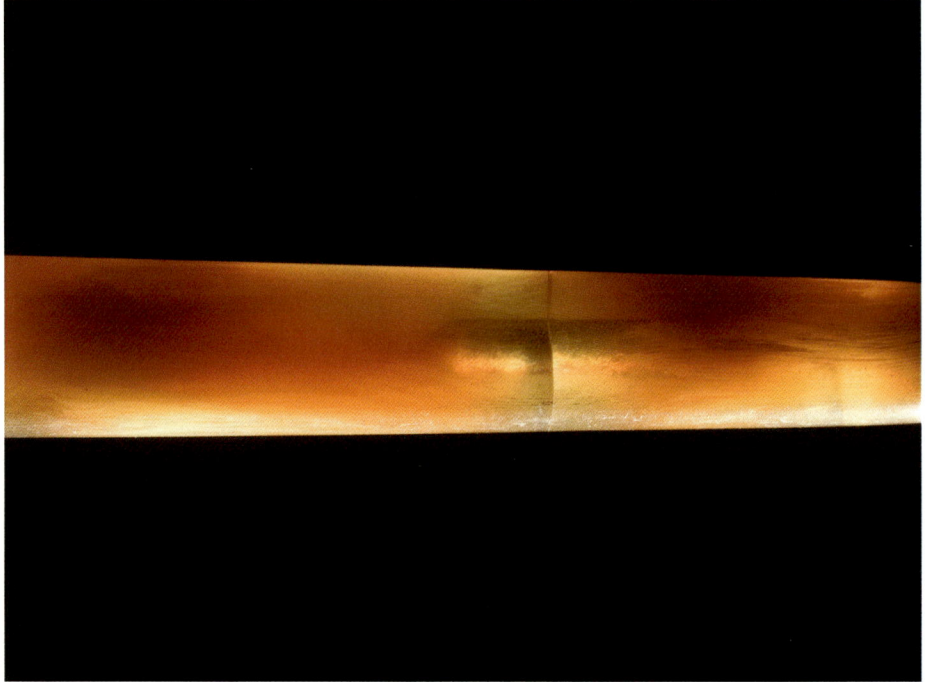

Figure 6.18
The corrosion layer formed on the iron-based shaft inside the cellulose nitrate handle of a butter knife occupies a greater volume than the metal alone and has caused cracking and bursting of the plastic.

6.4.2 *Degradation of cellulose acetate*

The majority of investigations on the degradation of cellulose acetate have been conducted on photographic film (cellulose triacetate) rather than moulded material. Like cellulose nitrate, cellulose acetate (CA) is deteriorated by both physical and chemical factors and the physical cause of degradation is plasticizer loss. Three-dimensional objects moulded from cellulose acetate comprise 20–40 per cent by weight plasticizer. Typical plasticizers include triphenyl phosphate (TPP), a solid with melting point 48.5°C (Wilson, 1995) which also acts as a flame retardant and is often used together with phthalate plasticizers such as dimethyl phthalate and DEHP (Hamrang, 1994). Migration and subsequent evaporation of plasticizer from between the cellulose acetate chains give rise to shrinkage, tackiness and increased brittleness (Figure 6.19).

Figure 6.19

A pile of transparent cellulose acetate tracing sheets from the 1980s illustrates typical physical deterioration. Sheets have shrunk and droplets of plasticizer can be seen at the surfaces.

The major chemical degradation reaction of cellulose triacetate is also similar to that of cellulose nitrate, the primary reaction being hydrolysis, also known as deacetylation, during which hydroxyl groups replace acetate groups (CH_3COO) on the cellulose ring, with the evolution of acetic acid (CH_3COOH) (Figure 6.20). Studies using nuclear magnetic resonance have shown that the loss of acetate groups is not random but occurs first from the

Figure 6.20

Deacetylation of
cellulose acetate
showing one repeat
unit. There are n repeat
units in a cellulose
acetate polymer.

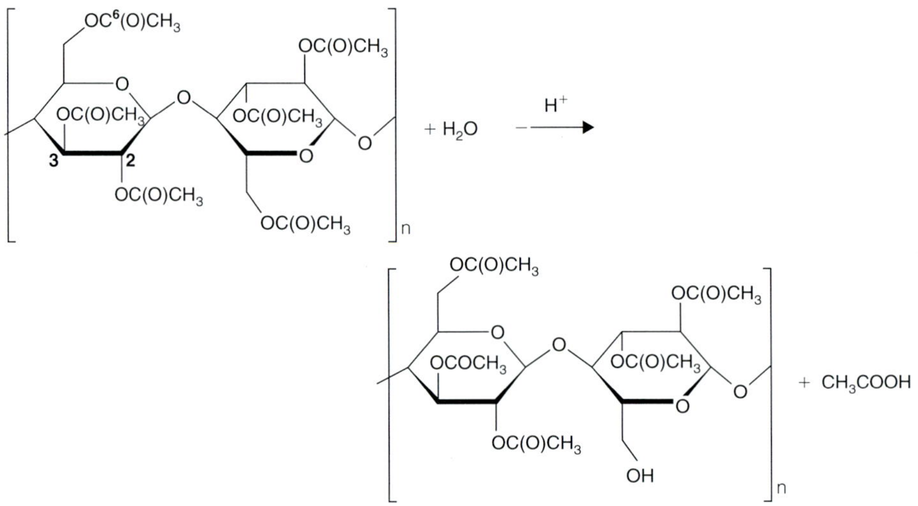

carbon at the 2 position on the ring, then from the 6 position and finally from
the carbon at position 3 (Derham et al., 1992). This is unexpected since carbons at the 2 and 3 positions are both secondary alcohols while carbon at position 6 is primary, but has been attributed to the accessibility of reactive groups
in the original cellulose from which cellulose acetate was manufactured.
Cellulose acetate undergoes autocatalytic breakdown if acetic acid is allowed
to remain in contact with the degrading polymer. This happens easily because
the solubility of acetic acid in cellulose acetate is high, similar to the solubility
of acetic acid in water in atmospheric moisture (Ligterink, 2002).

Deacetylation is accelerated by water (usually in the form of moisture in
air), acid or base. Because the loss of acetyl groups from cellulose acetate
results in the formation of acetic acid (CH_3COOH), which gives a distinct
vinegar-like odour to degrading materials, the process is also known as the
vinegar syndrome. Because the acidic vapours are mobile, acetic acid produced
by deteriorating cellulose acetate can catalyze degradation of other organic
materials in the vicinity (Figure 6.21). This phenomenon has been particularly detected in a brand of dolls known as Pedigree and has therefore been
described as 'Pedigree Doll's Disease' (Edwards et al., 1993). Metals in the
vicinity are corroded by the acetic acid produced by deacetylation. With time
and loss of acetate groups, the production of acetic acid lessens and the cellulose acetate is reduced to cellulose (Figure 6.22).

The degradation of TPP has also been shown by some to increase acidity of
cellulose acetate (Shinagawa et al., 1992), although others found no such con-
nection (Ram, 1990). Triphenyl phosphate, used as a plasticizer for cellulose
acetate since the 1940s, decomposes to form diphenyl phosphate and phenol.
Diphenyl phosphate is a strong acid so is likely to accelerate deacetylation of

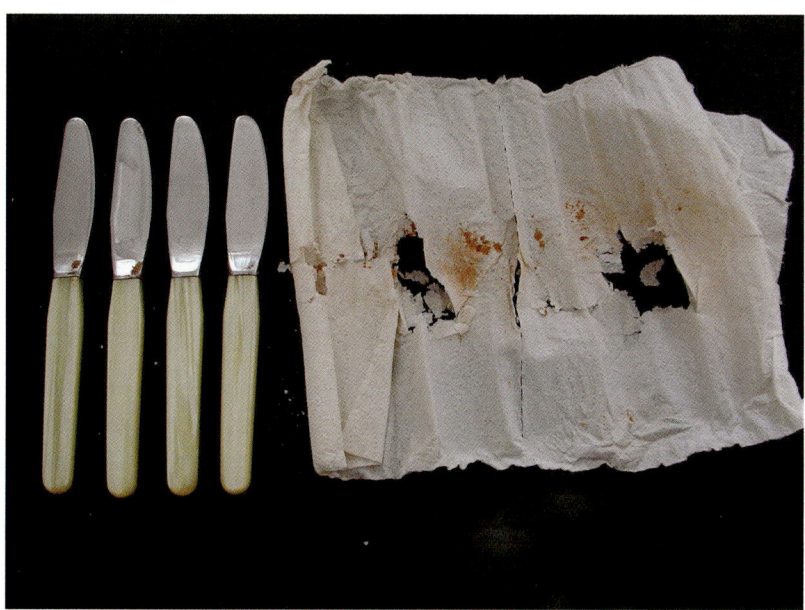

Figure 6.21

Acetic acid produced by degrading knife handles from the 1950s has attacked the paper in which the knives have been wrapped during storage. The knife blades have also undergone corrosion on contact with acetic acid.

Figure 6.22

Two cellulose acetate spoon handles from the 1950s. Acid Detection (A-D) strips enclosed with the spoons indicate that the spoon handle in good condition (left) produces a high concentration of acetic acid, while the highly deteriorated example (right) produces none. The handle on the right is almost fully deacetylated and chemically similar to cellulose.

the cellulose acetate polymer. Residues of sulphuric acid used as a catalyst in the synthesis of cellulose acetate are also thought to accelerate degradation.

Although deacetylation is the primary degradation pathway, chain scission takes place simultaneously, although the activation energy required to break bonds between cellulose rings is greater than for deacetylation (Ram, 1990). Chain scission results in a decrease in cellulose acetate's molecular weight and an increase in its solubility. Thermal oxidation of the hydroxyl groups, formed at the cellulose rings by deacetylation, has been proposed (Ram and McCrea, 1988). The carbonyl groups in the ketones and aldehydes formed on oxidation are chromophores and therefore blamed for yellowing of cellulose triacetate films.

6.4.3   Degradation of plasticized PVC

In many museum collections degradation of plasticized PVC materials, such as clothing and footwear, furniture, electrical insulation, medical equipment, housewares, toys and packaging materials used to store objects, has been detected as little as 5–10 years after acquisition. Degradation is usually manifested first as tackiness at surfaces accompanied by a glossy appearance, then by discolouration and sometimes as white crystals at surfaces. The degradation pathways exhibited by plasticized PVC are the result of degradation of the two major components of its formulation, namely polymer and plasticizer. Although the two components deteriorate independently of each other, the resulting products destabilize the whole.

Plasticizers are the major modifier for PVC formulations in terms of percentage weight (between 15 per cent for vinyl flooring and 50 per cent for waterproof boots) and physical properties. A plasticizer is a semi-volatile solvent material incorporated into a polymer or polymer mixture to increase its workability, flexibility and elongation (Wilson, 1995). Plasticizer molecules are evenly dispersed throughout the PVC and attach themselves weakly to the surfaces of polymer chains via Van der Waals interactions. The largest single product used as a general purpose plasticizer worldwide since the 1950s is di (2-ethylhexyl) phthalate (DEHP), which has been identified in many museum plastics (Shashoua, 2001). The tackiness and development of high gloss in deteriorating PVC indicate that liquid plasticizer is present at surfaces (Figure 6.23). From there, plasticizer evaporates at a rate dependant on its vapour pressure. DEHP has a boiling point of 386°C and evaporates slowly under ambient conditions.

PVC is susceptible to degradation when exposed to heat and light. The rate of degradation is greater in the presence of oxygen. The pathway by which degradation of the PVC polymer takes place is complex. In general, it comprises one major reaction, namely the evolution of hydrogen chloride (dehydrochlorination). In addition, crosslinking and chain scission reactions affect

Figure 6.23

Two life size crash test dummies, used to demonstrate the importance of wearing seat belts in moving cars in the 1970s, are constructed from plasticized PVC. Their surfaces are shiny and discoloured and their forms have softened and distorted due to the migration of plasticizer. Glass dishes have been placed in the well-sealed showcase to collect dripping plasticizer.

the physical properties of the degraded PVC. Crosslinking results in high molecular weight and stiff polymers, while chain scission reduces the molecular weight, thereby increasing solubility.

The instability of PVC is due to its structural irregularities, which include carbon-carbon double bonds at chain ends, tertiary chlorides, oxygen-containing structures and catalyst residues from the polymerization process. These have lower thermal stability than the linked vinyl chloride units and, therefore, act as initiation sites for degradation even at very low concentrations. Dehydrochlorination occurs at imperfections in the PVC structure and starts with the breaking of a C—Cl bond (Figure 6.24). Loss of a chlorine atom is

$$-CH_2-\underset{\underset{Cl}{|}}{CH}-CH_2-\underset{\underset{Cl}{|}}{CH}- \longrightarrow -CH_2-CH-CH_2-\underset{\underset{Cl}{|}}{CH}- \quad + \quad -Cl$$

$$\downarrow$$

$$-CH_2-CH=CH-CH_2-\underset{\underset{Cl}{|}}{} \quad + \quad HCl$$

$$\downarrow \textit{'unzipping' process}$$

$$\sim CH=CH-CH=CH-CH=CH-CH=CH\sim$$

$$+ \text{ hydrogen chloride}$$

Figure 6.24

Degradation of PVC polymer results in the formation of polyenes and hydrogen chloride, via an unzipping pathway.

followed almost immediately by abstraction of a hydrogen atom and a shift of electrons in the polymer to form a double bond. The next chlorine becomes allylic, highly reactive and is readily removed. This leads to the progressive 'unzipping' of neighbouring chorine and hydrogen atoms to form a conjugated polyene system (alternate single and double carbon bonds), accompanied by the formation of hydrogen chloride. As the conjugated polyene system develops, the polymer begins to absorb radiation in the ultraviolet part of the spectrum.

After between 7–11 repeat polyene units have formed, absorption of light shifts to longer wavelengths until the deteriorated PVC is absorbing in the violet, blue and green parts of the spectrum. The rate of degradation can be followed using colour changes from white to yellow to orange to red, brown and, ultimately black. Dehydrochlorination is an autocatalytic reaction so if the hydrogen chloride produced is not removed from the environment surrounding PVC, degradation continues at an accelerated rate. Dehydrochlorination is greatly enhanced in the presence of oxygen, particularly at the sites of double bonds. Acids, ketones and carbon-carbon unsaturation are formed in addition to hydrogen chloride. Chain scission occurs more readily than crosslinking between two adjacent chains.

The rate and extent of degradation of the PVC polymer and the migration and loss of plasticizer, particularly phthalates, are related. Addition of phthalate plasticizers to PVC has been shown to reduce the rate of dehydrochlorination by the polymer, by inhibiting the growth of polyene sequence (Beltran and Marcilla, 1997). DEHP inhibits the degradation of the PVC polymer, therefore when it either migrates to surfaces or is absorbed by other materials, PVC discolours, becoming tacky to the touch and embrittle.

Like all esters, phthalate plasticizers are susceptible to hydrolysis (addition of water) when exposed to strongly acidic or alkaline environments. Boiling DEHP for 1 hour at 386°C in alkaline conditions results in a 30 per cent decomposition into phthalic anhydride, 2-ethyl hex-1-ene and 2-ethylhexanol (Wilson, 1995). Because the latter two products are volatile, they are not seen, but phthalic anhydride is white and crystalline. Acid conditions lower the temperature of decomposition and result in the formation of phthalic acid, also a white crystalline solid (Figure 6.25). Acidic conditions develop when PVC polymers undergo dehydrochlorination and form acidic hydrogen chloride. Although phthalic anhydride and acid disfigure objects, they do not accelerate degradation (Figure 6.26).

Oxygen attack on alkyl groups in the DEHP molecule also results in the formation of phthalic acid. There are two chemically identical tertiary C—H positions that are susceptible to oxidation (Figure 6.27). Oxidation of DEHP may not be detected visually until it reaches an advanced stage, causing discolouration to occur and unpleasant odours to develop. It is attributed to overheating during processing.

di-2-ethylhexyl phthalate

$$CH_3$$
$$|$$
$$CH_2$$
$$|$$
CO—O—CH$_2$—CH—CH$_2$—CH$_2$—CH$_2$—CH$_3$

CO—O—CH$_2$—CH—CH$_2$—CH$_2$—CH$_2$—CH$_3$
$$|$$
$$CH_2$$
$$|$$
$$CH_3$$

alkali ↙   ↘ *acid*

phthalic anhydride   phthalic acid

.CO.   .COOH
   O
.CO'   .COOH

+

2-ethylhex-1-ene

$$CH_3$$
$$|$$
$$CH_2$$
$$|$$
CH$_2$=C—CH$_2$—CH$_2$—CH$_2$—CH$_3$

+

2-ethylhexanol

HO—CH$_2$—CH—CH$_2$—CH$_2$—CH$_2$—CH$_3$
$$|$$
$$CH_2$$
$$|$$
$$CH_3$$

Figure 6.25

DEHP hydrolyzes to form white, crystalline phthalic acid under acidic conditions. Acid conditions arise from dehydrochlorination of the PVC polymer.

Figure 6.26

A cockled and distorted PVC photograph pocket has lost plasticizer after 30 years in use (left). Crystals of phthalic acid are visible on the upper surface of the positive image due to hydrolysis of DEHP (right).

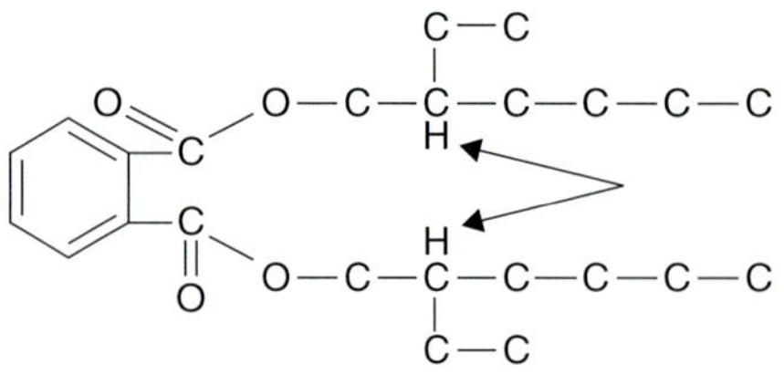

Figure 6.27

Two equivalent tertiary C—H positions in the DEHP molecule make it vulnerable to oxidation.

6.4.4   Degradation of polyurethane foam

Polyurethane foams deteriorate via photo-oxidation, thermal oxidation and hydrolysis. The many pores or cells in polyurethane foams make the polymer highly accessible to oxygen, light and water in air. Their physical structure causes a higher rate of degradation than films, fibres or cast blocks of an equivalent formulation. In addition, the processing of foams may involve blowing air through polyurethanes in liquid form, providing perfect conditions for oxidation. The extent of sensitivity of polyurethane foams to chemical degradation factors is dependant on the polyol base used. Polyester-based polyurethanes are more readily hydrolyzed than polyether-based materials. Polyester-based polyurethanes are hydrolyzed under alkali conditions resulting in saponification of the ester group and formation of carbonyl and alcohol groups (Pavlova et al., 1985). The degradation products catalyze further hydrolysis. By contrast, polyether-based polyurethanes are subject to acid hydrolysis.

Polyurethane ether (PUR-ether) foams are thought to degrade primarily by oxidation, particularly in the presence of light, resulting in discolouration and a loss of mechanical properties. Polyurethanes synthesized from a polyether polyol and an aromatic diisocyanate such as diphenylmethane diisocyanate (MDI) are highly vulnerable to photo-oxidation, whereas polyester-based polyurethanes are more resistant to ultraviolet radiation (Kerr and Batcheller, 1993). Metal ions, particularly copper, aluminium and zinc, form chelates with some polyurethanes imparting increased sensitivity to photo-oxidation (Rånby and Rabek, 1975).

Photo-oxidation results in chain scission, in which energy breaks polymer chain bonds to create a polymer with two or more shorter chains and is manifested by collapse and crumbling of foams. Scission of the urethane link (—NHCOO—) results in the formation of amino and carbonyl groups with the evolution of carbon monoxide and dioxide. Crumbling often starts at surface skins of foams and, when the surfaces crumble and fall away from the object, fresh undegraded foam is exposed to light, perpetuating degradation to the point of complete destruction (Figure 6.28). It is possible to follow the progress of degradation from the depth of crumbling (Keneghan, 1999). Degradation of uncoated polyurethane foam is visible after around 20 years whereas Piero Gilardi's *Still Life of Watermelons*, a sculpture constructed from

Figure 6.28
Photo-oxidation of polyurethane ether foam frequently results in loss of structure and crumbling.

painted PUR-ether in the mid 1960s, is still in good condition after 30 years due to protection of the polymer from light (van Oosten, 1999).

6.5   Terms used to describe the degradation of plastics objects

Visual examination of plastics to assess their level of degradation is usually the first stage in planning the resources necessary to store, conserve and display collections. There are several approaches to making a condition survey of a large collection containing plastics. The methodology used by the British Museum, V&A Museum, and the National Museum of Science and Industry (Science Museum) in the United Kingdom and the National Museum of Denmark, has been to survey collections in their original storage or display areas rather than relocating them for examination. Because the collections comprised more than 5000 objects and there were insufficient resources to examine all of them, every tenth object containing plastic was randomly selected for detailed assessment (Shashoua and Ward, 1995). Objects were examined for visible signs of degradation, while types of degradation were attributed, where possible, to physical, chemical or other factors. Information about storage materials and the microclimate was also recorded to ascertain whether there was a relation between condition of the object and storage conditions. Because one of the purposes of the survey was to extract statistical information, the survey form used was transferred to an electronic database. If a degradation profile of plastics with time is to be built up, condition surveys should be carried out at regular intervals.

The scoring system for assessing the condition of objects and requirements for conservation varies between institutions. It is most often a four-point scale where one end indicates that the object is in excellent condition and requires no conservation while the other end suggests that the material is actively deteriorating and needs treatment urgently. The terms to describe degradation used in condition surveys of plastics in the United Kingdom are based on those developed by the plastics industry to define faults in plastics products (Then, 1996). Because it is complex to translate written definitions accurately into other languages, I propose additional illustrated examples of each term in Appendix 3.

References

Beltran, M. and Marcilla, A. (1997). PVC plastisols decomposition by FTIR spectroscopy. *European Polymer Journal*, **33** (8), 1271–1280.

Beyler, C. L. and Hirschler, M. M. (2002). Thermal decomposition of polymers. In *SFPE Handbook of Fire Protection Engineering*, 3rd edition. (P. J. DiNemmo, ed.) pp. 1/110–1/131. NFPA, Quincy, MA, USA.

Bradley, S. B. (1994). Do objects have a finite lifetime?. In *Care of Collections* (S. Knell, ed.) pp. 51–59. Routledge.

Brydson, J. A. (1999). *Plastics Materials*, 6th edition. Butterworth-Heinemann.

Cappitelli, F., Zanardini, E. and Sorlini, C. (2004). The biodegradation of synthetic resins used in conservation. *Macromolecular Bioscience*, **4**, 399–406.

Derham, M., Edge, M., Williams, D. A. R., Williamson, D. M. (1992). The Degradation of cellulose triacetate studied by nuclear resonance spectroscopy and molecular modeling. In: *Postprints of Polymers in Conservation conference* Manchester, 17–19 July 1991 (N.S. Allen, M. Edge and C.V. Horie, eds.) pp. 125–137 Royal Society of Chemistry.

Edwards, H. G. M., Johnson, A. F., Lewis, I. R. and Turner, P. (1993). Raman spectroscopic studies of Pedigree Doll disease. *Polymer Degradation and Stability*, **41**, 257–264.

Feller, R. L. (1977). Stages in the degradation of organic materials. Preservation of paper and textiles of historic value. In *Advances in chemistry series, no.* (J. C. Williams, ed.) pp. 314–335. American Chemical Society, Washington D. C.

Grassie, N. (1966). Fundamental Chemistry of Polymer Degradation. In *Weathering and Degradation of Plastics* (S. H. Pinner, ed.) pp. 1–13. Columbine Press, London.

Grattan, D. W. (1993). Degradation rates for some historic polymers and the potential of various conservation measures for minimizing oxidative degradation. In: *Postprints of Saving the Twentieth Century: The Conservation of Modern Materials* Ottawa, 15–20 September 1991 (D. Grattan, ed.) pp. 351–361. Canadian Conservation Institute.

Hamrang, A. (1994).Degradation and stabilisation of cellulose based plastics and artefacts, Ph.D. thesis, Manchester Metropolitan University.

Keneghan, B. (1999). Here Today, Gone Tomorrow? Problems with plastics in contemporary art. In: *Postprints of Modern Art: Who Cares?* 8–10 September 1997 (U. Hummelen and D. Sillé, eds) pp. 356–361. The Foundation for the Conservation of Modern Art and the Netherlands Institute for Cultural Heritage, Amsterdam.

Kerr, N. and Batcheller, J. (1993). Degradation of polyurethanes in 20th-century museum textiles. In: *Postprints of Saving the Twentieth Century: The Conservation of Modern Materials* Ottawa, 15–20 September 1991 (D. Grattan, ed.), pp. 189–206. Canadian Conservation Institute.

Ligterink, L. J. (2002). Notes on the use of acid absorbents in storage of cellulose acetate-based materials. In *Contributions to Conservation, Research in Conservation at the Netherlands Institute for Cultural Heritage* (J. A. Mosk and N. H. Tennent, eds) pp. 64–73. James and James.

McNeill, I. C. (1992). Fundamental aspects of polymer degradation. In: *Postprints of Polymers in Conservation* Manchester 17–19 July 1991 (N. S. Allen, M. Edge and C. V. Horie, eds.) pp.14–31. Royal Society of Chemistry.

Mecklenburg, M. F., McCormick-Goodhart, M. and Tumosa, C. S. (1994). Investigation into the degradation of paintings and photographs using computerized modelling of stress development. *Journal of the American Institute for Conservation* [online] **33** (2) pp.153–170. Available from: http://aic.stanford.edu/jaic/articles/jaic33-02-007_3.html [Accessed 2 April 2006].

Michalski, S. (2002). Double the life for each five-degree drop, more than double the life for each halving of relative humidity. In: *Preprints of the 13th ICOM-CC Triennial Meeting*, Rio de Janeiro, 22–27 September 2002 (R. Vontobel, ed.) pp. 66–72). James & James Ltd.

Mills, J. S. and White, R. (1994). *The organic chemistry of museum objects*, 2nd edition. Butterworth-Heinemann.

Moriyama, Y., Naoko, K., Inoue, K. and Kawaguchi, A. (1993). *International Biodegradation and Biodegradation*, **31**, 231.

Nason, K., Carswell, T. S. and Adams, C. H. (1951). Low temperature behavior of plastics. *Modern Plastics*, (December), 127–203.

Pagliarino, A. (1999). Plain Plastics. *AICCM National Newsletter*, **71**, 1–8.

Parkins, J. A. (1957). Behaviour of cellulose nitrate and finishes in light. *Paint and Varnish production*, , 42–44.

Pavlova, M., Oraganova, M. and Novakov, P. (1985). Hydrolytic stability and protective properties of polyurethaner oligomers based on polyester/ether/polyols. *Polymer*, **26** (11), 1901–1905.

Quackenbos, H. M. (1954). Plasticizers in vinyl chloride resins. *Industrial and Engineering Chemistry*, **46** (6), 1335–1341.

Rånby, B. and Rabek, F. F. (1975). *Photodegradation, photo-oxidation and photostabilization of polymers*. John Wiley, London.

Ram, A. T. (1990). Archival Preservation of Photographic Films – a perspective. *Polymer Degradation and Stability*, **29**, 3–29.

Ram, A. T. and McCrea, J. L. (1988). Stability of processed cellulose ester photographic films. *SMPTE Journal*, **97**, 474–483.

Rice, P. and Adam, H. (1977). *Developments in PVC production and processing*. (A. Whelan and J. L. Craft, eds) Applied Science Publishers, London.

Rivers, S. and Umney, N. (2003). *Conservation of furniture*. Butterworth-Heinemann.

Rychly, J. and Strlic, M. (2005). Degradation and ageing of polymers. In *Ageing and Stabilisation of Paper* (M. Strlic and J. Kolar, eds) pp. 9–23. National and University Library, Ljubljana, Slovenia.

Sears, J. K. and Darby, J. R. (1982). *The technology of plasticizers*. John Wiley and Sons, New York.

Selwitz, C. (1988). *Cellulose Nitrate in Conservation*. The Getty Conservation Institute.

Shashoua, Y. (2001). Inhibiting the degradation of plasticized poly(vinyl chloride) – a museum perspective. Ph.D. thesis, Danish Polymer Centre, Technical University of Denmark.

Shashoua, Y. (2004). Modern Plastics – do they suffer from the Cold. In: *Preprints of IIC Congress Modern Art*, New Museums Bilbao, 13–17 September 2004 (A. Roy, P. Smith eds.) pp. 91–9. IIC.

Shashoua, Y. and Ward, C. (1995). Plastics:modern resins with ageing problems'. In *Preprints of Resins, Ancient and Modern* (M. M. Wright and J. H. Townsend, eds) pp. 33–37. SSCR, Aberdeen.

Shashoua, Y., Bradley, S. M. and Daniels, V. D. (1992). Degradation of cellulose nitrate adhesives. *Studies in Conservation*, **37**, 113–119.

Shinagawa, Y., Murayama, M. and Sakaino, Y. (1992). Investigation of the archival stability of cellulose triacetate film: The effect of additives to CTA support. In: *Postprints of Polymers in Conservation* Manchester 17–19 July 1991 (N. S. Allen, M. Edge and C. V. Horie eds.) pp. 138–150. Royal Society of Chemistry.

Silva, M. (2006). Investigating cellulose nitrate degradation caused by fungal attack. In: *Postprints of The Future of the Twentieth Century–collecting, interpreting and conserving modern materials*. Second Annual Conference Arts and Humanities Research Board Winchester 26–28 July 2005 (P. Garside, ed.) pp. 72–76. Archetype Publications. London.

Stewart, R., Littlejohn, D., Pethrick, R. A., Tennent, N. H. and Quye, A. (1996). The use of accelerated ageing tests for studying the degradation of cellulose nitrate. In: *Preprints of the 11th ICOM-CC Triennial Meeting* The Hague, 12–16 September 2005 (J. Bridgland, ed.) pp. 67–970. James&James Ltd.

Then, E. T. H. (1996). The care of a plastics collection. *Polymer Preprints, American Chemical Society, Division of Polymer Chemistry*, **37** (2), 166–167.

Then, E. and Oakley, V. (1993). A survey of plastic objects at the Victoria and Albert Museum. *V & A Conservation Journal*, **6**, 11–14.

van Oosten, T. (1999). Here Today, Gone Tomorrow? Problems with plastics in contemporary art. In: *Postprints of Modern Art: Who Cares?* 8–10 September 1997 (U. Hummelen and D. Sillé, eds) p. 162. The Foundation for the Conservation of Modern Art and the Netherlands Institute for Cultural Heritage, Amsterdam.

Weast, R. C. (1974). *Handbook of Chemistry and Physics*, 55th edition.. CRC Press.

Wiles, D. M. (1993). Changes in polymeric materials with time. In: *Postprints of Saving the Twentieth Century: The Conservation of Modern Materials* Ottawa, 15–20 September 1991 (D. Grattan, ed.) pp. 105–112. Canadian Conservation Institute.

Williams, R. S. (1993). Composition implications of plastic artefacts: a survey of additives and their effects on the longevity of plastics. In: *Postprints of Saving the Twentieth Century: The Conservation of Modern Materials* Ottawa, 15–20 September 1991 (D. Grattan, ed.) pp.135–153. Canadian Conservation Institute.

Williams, R. S. (2002). Care of Plastics: Malignant plastics. *WAAC Newsletter*, (January), **24**, 1. Available from: http://www.palimpsest.stanford.edu/waac/wn/wn24-1/wn24-102.html [Accessed 30 June 2006].

Wilson, A. S. (1995). *Plasticisers: Principles and Practice*. The Institute of Materials.

Conservation of plastics

Summary

Chapter 7 describes the options that are used today for slowing the rate of degradation of plastics (inhibitive conservation) and for invasively stabilizing and strengthening them (interventive conservation). The efficiency of some adsorbents used in inhibitive conservation is discussed, together with some simple techniques to monitor the rate of degradation of plastics. The options for interventive conservation of plastics in cultural collections have always been restricted by the need to adhere to codes of ethics for conservation professionals. It is difficult to reverse an adhesive repair or to remove a consolidant or protective coating from plastic without damaging the original surfaces. However, interventive treatments are used by dealers and insurance companies to maintain commercial value of the artwork and are being employed more widely by museum conservators.

Conservators consider two approaches to conservation when planning treatment to extend the useful lifetime of cultural materials: preventive or passive and active or interventive. Preventive conservation may be defined as limiting the rate and extent of degradation of materials by developing and implementing procedures for storage, exhibition, packing and transporting. Preventive conservation is an ongoing process that continues throughout the life of objects and does not end with interventive treatment (American Institute for Conservation of Artistic and Historic Works, 1997).

Once initiated, degradation of plastics cannot be prevented, reversed or stopped, but only inhibited or slowed. I prefer to use the term 'inhibitive' conservation in place of 'preventive'. Inhibitive conservation of plastics can either involve the removal or reduction of factors causing degradation including light, oxygen, acids and relative humidity, or of any breakdown products which accelerate degradation. If applied successfully, inhibitive conservation can help

Figure 7.1

Effect of successful
inhibitive conservation
on rate of degradation
of plastics.

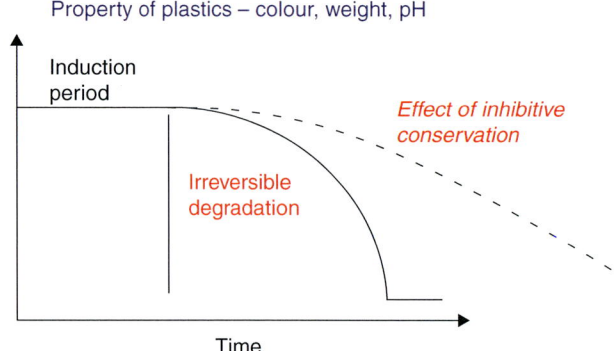

Figure 7.1

Effect of successful inhibitive conservation on rate of degradation of plastics.

to prolong the useful lifetime of many objects simultaneously, so is considered a highly effective use of resources (Figure 7.1).

Interventive conservation treatments are those involving practical, invasive treatments applied as necessary to objects to limit further degradation and often to preserve their significance. Treatments include adhering broken sections, cleaning surfaces and filling missing areas to strengthen objects weakened by degradation. Although most condition surveys of plastics in museums conclude that approximately 75 per cent of collections require cleaning, because dirt either affects their stability, chemical and physical properties or their significance, such interventive conservation practices are still poorly developed. The major reason for underdevelopment is the high sensitivity of many plastics, especially when degraded, to organic liquids, aqueous solutions and water itself. In addition, coatings or adhesives which adhere successfully to plastics' surfaces must soften the substrate either by dissolution, etching or melting to achieve bonding. Such treatments frequently change the appearance of the original, which may be unacceptable, particularly for works of art.

This chapter will first present the options which are widely used today for slowing the rate of degradation of plastics (inhibitive conservation) and then those for invasively stabilizing and strengthening them (interventive conservation). Interventive treatments are used by private dealers, collectors and art insurance companies to maintain the commercial value and original appearance of plastics artworks. Some interventive treatments are considered restoration rather than conservation.

7.1   Inhibitive conservation

All plastics are degraded to varying extents by exposure to ultraviolet and visible radiation, heat, oxygen and water, most commonly in the form of water vapour in air. Some pollutants act as degradation agents, whether they occur in the surrounding air, are offgassed by storage and display materials, present as residues of manufacture (for example, excess, unreacted monomer) or produced as degradation products by the plastic itself. Minimizing exposure to degradation factors and providing stable environmental conditions will slow the rate of breakdown.

Currently, there are no internationally accepted guidelines defining appropriate storage environments of plastics in museum in cultural collections as a generic group. Traditionally, those designed to preserve other fragile organic materials, such as feathers and plant fibres, have been applied to plastics (Bradley, 1990). These include maintaining a stable relative humidity, usually 55 ± 3 per cent, a temperature of 18 ± 2°C, maximum illumination between 50–300 lux, elimination of ultraviolet radiation and good ventilation. Although there are no publications which detail systematic research into appropriate storage environments for all plastics, investigations into individual plastic types suggest that those which deteriorate by hydrolysis require a lower storage relative humidity (RH) than those which are less sensitive to water. A relative humidity of 30 per cent has been proposed for polyether polyurethane foam, which degrades by hydrolysis (Lovett and Eastop, 2004). By contrast, plastics which are plasticized by water, such as casein-formaldehyde, require RH higher than 40 per cent if they are to avoid cracking due to dehydration.

In March 2006, the Danish National Cultural Heritage Agency (Kulturarvstyrelsen) issued draft guidelines for conserving and handling all materials including plastics in Danish museums. The proposal for long-term storage of cellulose nitrate (CN) and cellulose acetate (CA) plastics comprised a maximum illumination 50 lux, maximum ultraviolet 75 μW / lumen, temperature 2–5°C and 20–30 per cent RH while synthetics and casein were expected to withstand higher temperatures and to require higher RH partly to minimize the effects of electrostatic charge (5–25°C and 50–60 per cent RH) (Danish National Cultural Heritage Agency, 2006). The importance of controlling both temperature and relative humidity has been internationally recognized for the long-term storage of composite materials which contain plastics in thin layers, including magnetic carriers, optical media and photographic materials (ISO-11799:2003 and ISO 18925:2002). Recommendations include maximum light levels of 50 lux, temperature −5°C and 30–40 per cent RH for moving image colour film, and + 20°C, 40 per cent RH for optical media. Because some degrading plastics, particularly semi-synthetics, produce degradation products which corrode other organic materials and metals, they should be separated from cellulose, silver and iron (Quye, 1999).

In museum stores where objects are grouped by historical period and not by material type or condition, storage climates are based both on the average conditions required by all the materials in the location and what can be achieved with the resources available. Some plastics, particularly those which are actively degrading, benefit from microclimates achieved using adsorbents to lower the concentration of gases and vapours in the air around them. Others require separation from the climate in the store. Some plastics could be stored at low temperatures to prolong their useful lifetime. All three options will be discussed here.

7.1.1   Inhibition of degradation using adsorbents

Gas adsorbents, also known as molecular traps and scavengers, are used to inhibit the rate of degradation of plastics either by minimizing those factors initiating degradation or those implicated in autocatalysis, a phenomenon discussed in Chapter 6. Adsorbents may either be installed in an active filter system in showcases, storage areas or simply placed in petrie dishes or polyethylene bags inside storage boxes or enclosures.

Activated carbon, also known as activated charcoal or activated coal, is often used to reduce the rate of degradation of cellulose nitrate objects and has proven effective for ethnographic materials including mock tortoiseshell boxes, hair combs and shadow puppets (Ward and Shashoua, 1999). Activated carbon has a huge surface area of 300–2000 m^2 per gram. By comparison, silica gel, another widely used adsorbent, has an effective surface area of 2–5 m^2 per gram. Activated carbon can also be obtained in the form of woven textiles, impregnated card or paper and pellets of various sizes. Pellets and powder are likely to adhere to tacky and uneven surfaces, so should not be allowed to come into direct contact with deteriorating materials. One of the most widely used adsorbents for applications in the chemical industry (cleaning gases), the home (cooker hoods) and medicine (charcoal tablets), activated carbon is more effective at adsorbing large non-polar molecules such as volatile organic acids, alcohols and ketones than low boiling materials such as formaldehyde, hydrogen chloride and ammonia.

Molecules are transported from surfaces of activated carbon by diffusion and trapped on sites by physisorption, primarily via weak London dispersion forces. Activated carbon cloth or carbon-impregnated paper, used as packing materials for cellulose nitrate, readily adsorb nitrogen oxides – CN's primary degradation

Figure 7.2

Gaseous pollutants in air compete with nitrous oxide to be adsorbed by activated carbon. Such competition reduces the effectiveness of activated charcoal in removing nitrous oxide. If nitrous oxides are allowed to remain in contact with CN objects, autocatalysis leads to their rapid degradation.

products. The degradation products are thus unable to participate in autocatalytic breakdown or to come into contact with metals or other organic materials in the vicinity. When used as a filter in an active system, carbon has been shown to remove 90 per cent nitrous oxides and 95 per cent of sulphur dioxide from air in a single pass (Hatchfield, 2002). Because of the wide range of pollutants readily adsorbed by activated carbon, gaseous pollutants in air may compete with nitrogen oxides for sites (Figure 7.2). When all sites are occupied, no further adsorption is possible until some become vacant. Since activated carbon is not self-indicating, it is difficult to see that it is exhausted and no longer effective at inhibiting degradation of CN. It can be regenerated by heating to 650°C in an inert atmosphere, causing adsorbed material to be desorbed, but the low cost of active carbon allows frequent replacement with fresh absorbent.

Although widely used by conservators to inhibit the rate of degradation of CN, activated carbon may not be the most effective molecular trap for nitrogen oxides. Measurements using Drager tube gas detectors to determine the amount of nitrogen oxides adsorbed by seven molecular traps suggested that zeolites (hydrated silicates of calcium and aluminium with a porous honeycomb structure) and activated charcoal were equally effective at adsorbing nitrous oxides from air (Kessler, 2002). In medical applications, where zeolites are used to deliver nitrous oxide gas to human blood to prevent clotting, doping zeolites with positively charged ions has been shown to pull the gas into their pores in preference to water or pollutants (Davis, 2005). Such specificity in an adsorbent prolongs its effective lifetime so that frequent replacement is less necessary than is the case with activated carbon.

Zeolites comprise a class of hydrated silicates of metallic ions, most frequently calcium and aluminium, which contain micropores of predetermined diameters. The dimensions of the micropores determine which gases, vapours or liquids particular zeolites are able to trap. Surface areas of zeolites are up to several hundred square meters per gram, which allows them to adsorb gases and vapours up to 30 per cent of their own dry weight. Whereas activated carbon holds molecules weakly by physisorption, molecules are adsorbed onto the micropore walls of zeolites by chemisorption or covalent bonds, which contain about 10 times greater energy than physical bonds.

Zeolites were first shown to inhibit the rate of degradation of photographic negative film based on cellulose acetate in 1994 by adsorbing acetic acid (Ram et al., 1994). The technique has since been adapted to inhibit the degradation of three-dimensional materials containing cellulose acetate including handbags, jewellery and modern art (Derrick et al., 1993). It has been asserted that zeolites inhibit the degradation of cellulose acetate by both trapping water vapour from the polymer, thereby minimizing the rate of hydrolysis, also known as deacetylation, and by trapping acetic acid vapour, which reduces the opportunity for more dominant, rapid autocatalysis to occur (Ligterink, 2002).

When both water and acetic acid are present, they compete for sites in the zeolite lattice.

Movie films in archives are usually stored in metal cans or plastic containers to protect the films both from mechanical damage during handling and from fire. Zeolites, usually Type 4A, enclosed in polyethylene sachets are introduced to the container. In contrast, three-dimensional objects containing cellulose acetate are rarely enclosed in air-impermeable containers but are either stored in lidded cardboard boxes or on open shelves. The difference in storage techniques has no scientific basis so is likely to be a matter of tradition and convenience. The dilemma arises of whether to ventilate cellulose acetate to disperse acetic acid, perhaps inadvertently in the direction of other objects, or to enclose them with zeolites. While enclosures prevent contact between water vapour in air and cellulose acetate, they reduce visibility of objects to curators, students and conservators. Ventilation seems to be favoured for modern artworks containing cellulose acetate whereas ethnographic materials are more often enclosed with zeolites (Derrick et al., 1994).

Since the degradation of most synthetic plastics involves reaction with oxygen, removing oxygen could be expected to limit the extent of the reaction. However, to date, only the rates of crazing, crumbling and discolouration of natural rubber (Shashoua and Thomsen, 1993) have been shown in large-scale trials to be slowed by storage in an oxygen-free atmosphere. Packing objects in a nitrogen atmosphere is the traditional method to achieve oxygen-free storage, but its long-term effectiveness is dependent on using a perfect barrier to prevent ingress of oxygen from the surrounding air.

A more convenient option has been used since the 1990s. Commercial oxygen absorbers have been designed to inhibit the oxidation of foods during transport, and have been evaluated for their suitability for use with museum plastics (Grattan and Gilberg, 1994). Oxygen absorbers comprise gas-permeable plastic sachets containing finely-divided iron. The iron oxidizes to iron oxides in the presence of oxygen and water thus binding oxygen (Figure 7.3). Moisture is provided by the presence of hygroscopic potassium chloride in the sachet, which absorbs water and is also a by-product of the oxidation reaction. Oxygen absorbers have been used as a low- cost, convenient alternative to flushing with nitrogen for oxygen-free storage (Shashoua and Thomsen, 1993). It is claimed that oxygen absorber can reduce the oxygen concentration of an airtight container down to 0.01 per cent (100 ppm) or less (Conservation by Design, 2007).

Objects are placed in an oxygen-impermeable envelope, such as those prepared from transparent laminated films of nylon and polyethylene or ceramic-coated polymers, into which oxygen absorber sachets have been introduced. Envelopes are flushed with dry nitrogen to remove any oxygen before being heat-sealed. When the iron in an oxygen absorber reacts with oxygen, it occurs via an exothermic reaction producing a small amount of heat (Shashoua, 1999).

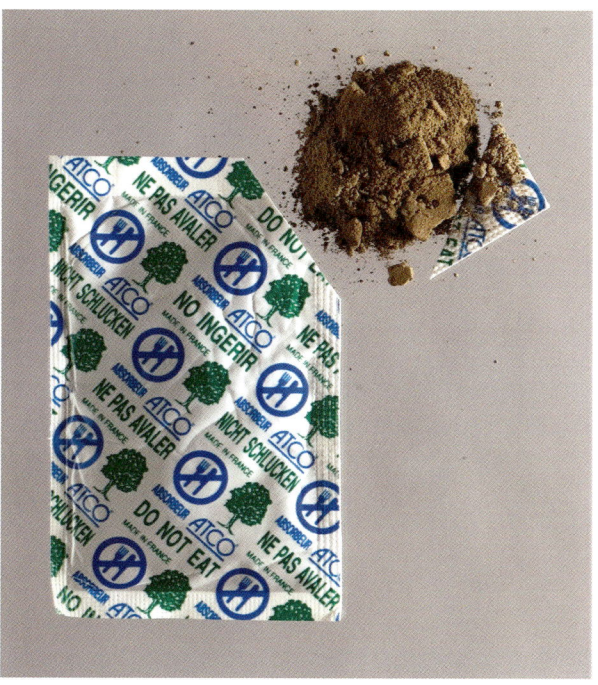

Figure 7.3

An oxygen-free microclimate can readily be achieved using oxygen absorbers, based on finely-divided iron, and developed for food packaging. As the fresh enclosed activated iron (left image) reacts with oxygen, it forms clumps of rust (right image).

In addition, as a by-product of this reaction, a small quantity of water is formed which causes the relative humidity to increase inside the enclosure.

Silica gel is widely used to buffer the relative humidity in storage areas and showcases, but can also adsorb degradation products from plastics. Although silica gel adsorbs nitrogen oxide and sulphur dioxide from air, both of which are degradation products of cellulose nitrate, it is only effective for around 24 hours (Parmar and Grosjean, 1991). The short-term effectiveness of silica gel can be attributed to its greater affinity for water molecules, which preferentially occupy the sites on its surfaces, thus limiting the sites available to pollutants. Silica gel can reduce the useful lifetime of PVC. It has proved effective at adsorbing phthalate plasticizer from PVC, thus reducing the polymer's resistance to discolouration by dehydrochlorination (Shashoua, 2001).

Few adsorbents are self-indicating and require external systems to show when they are exhausted or otherwise ineffective. Chemical indicator-impregnated string suspended over CN objects or an indicator outline drawn around objects on packaging materials have been used to highlight acidic emissions and act as early warning systems to renew activated charcoal or zeolites. Cresol red (o-cresolsulphponephthalein) and cresol purple (m-resolsulphponephthalein) change colour in contact with low concentrations of acidic vapours (Fenn, 1995). Cresol red is orange coloured at pH 0.2, yellow at pH 1.8 and reddish-purple at pH 8.8. Cresol purple is red at pH 1.2, yellow at pH 2.8 and purple at pH 9.0. They can be dissolved in water or ethanol and applied to paper for convenient use.

Universal pH indicator papers can be used to determine an increase in acidity (reduction in pH value) but, since they need to be wet prior to measuring, they are unsuitable for continuous monitoring. Glass sensors developed by the Fraunhofer Institute have been used to detect the presence of acidic gases including nitrogen oxides produced by cellulose nitrate (Fuchs and Leissner, 1995). Glass sensors are pre-corroded to a standardized level before exposing them. The presence of acidic gases is determined from the intensity of the corrosion induced on exposure. Unfortunately, the system cannot distinguish between different acids.

A-D (acid detection) strips were developed by the Image Permanence Institute (IPI) to identify degradation of cellulose acetate-based photographic film or negatives semi-quantitatively. However, they have also been used as indicators of active degradation of cellulose acetate objects (Figure 7.4). The active reagent in the strips is bromocresol green which changes from blue (at pH 5.4) to green (at pH values between 5.4–3.8), which indicates production of acetic acid. A-D strips become yellow below pH 3.8 suggesting that degradation is at an advanced stage and copying of the film or negative is imperative. The strips are placed in proximity to, but not touching, the material to be monitored for 24 hours and evaluated promptly because the colour soon reverts to blue after removal from the test enclosure. A second A-D strip should be stored away

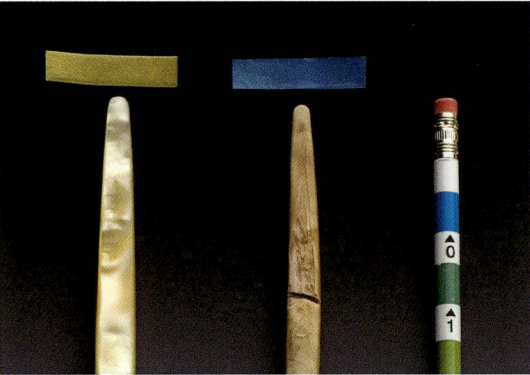

Figure 7.4

A-D strips have been used to indicate active degradation in cellulose acetate spoon handles from the 1950s. Compared with the scale provided with the strips (far right of image), more acetic acid is being produced by the handle on the left. This may be explained by the high level of degradation shown by the handle on the right, which resembles paper.

from the material under study to act as a control. A-D strips contain a humectant, probably glycerol, so that they do not require liquid water to activate them and can be used as long-term monitoring devices. However, RH below 30 per cent and temperature below 15°C extends the reaction time to 4 days.

Although developed for specific detection of acetic acid, A-D strips have been used to indicate the presence of other acidic gases, due to their convenience and ease of interpretation. I have used A-D strips to detect the onset of production of nitrogen oxides by cellulose nitrate, however the progressive colour changes in strips expected with increasing degradation of cellulose acetate did not take place with CN. Similar products to A-D strips include Danchek®, thought to comprise bromocresol green dispersed in silica gel (Hatchfield, 2002).

Oxygen indicators are often supplied with oxygen absorbers. They are in the form of a pressed tablet which changes colour from pale pink (less than 0.1 per cent oxygen) to dark blue (greater than 0.5 per cent). However, the active dye in indicators tends to lose its sensitivity to oxygen after approximately six months, manifested by unreliable colour changes, so an oxygen monitoring device is more reliable. An alternative method to using sachets of oxygen absorber is to enclose objects in a multilayer polymer film, which incorporates the oxygen scavenger agents as one of its layers. Such films require exposure to ultraviolet radiation to initiate the scavenging reaction (Butler, 2006).

7.1.2   Inhibition of degradation by enclosing plastics

Both observation of naturally aged plasticized PVC in collections and accelerated ageing of model materials representing such objects when newly acquired suggest that the useful lifetime of the PVC polymer and the migration of plasticizers are related (Shashoua, 2001). Di-2-ethylhexyl phthalate (DEHP), the most commonly identified plasticizer in museum objects containing PVC, inhibits the degradation of the PVC polymer. When it is lost, PVC becomes discoloured and brittle. Weight loss has been used to quantify loss of plasticizer, infrared spectroscopy developed to quantify concentration of DEHP at surfaces, and optical densitometry used to quantify darkening of samples.

The goal of inhibitive conservation in the case of plasticized PVC is to prevent loss of plasticizer, thereby protecting the PVC polymer from dehydrochlorination. Degradation of both PVC model materials and naturally degraded objects has been slowed by enclosing them in non-adsorbent media such as glass and polyester envelopes. Storage in open environments with circulating air, such as shelves, or wrapping objects in plasticizer-absorbing materials such as polyethylene results in loss of between 50–60 per cent of the initial DEHP within three months under ambient conditions (Shashoua, 2001). Low density polyethylene readily absorbs oily materials, including plasticizers, as illustrated by a polyethylene fishing box used to store highly plasticized PVC lures for

Figure 7.5

A polyethylene fishing box has absorbed plasticizer from PVC lures, as seen in the furthest right compartment, and formed polyethylene 'fossils' over a 20-year period.

20 years (Figure 7.5). With time, DEHP diffused from the lures into the poly-ethylene, softening the box so that it has deformed to the shape of the lures and today resembles plastic fossils. The lures became brittle and subsequently broke. Traditionally, low density polyethylene bags have been used to store many types of objects. They are clearly unsuitable for storing plasticized PVC.

Calculations based on the time taken for model sheets to lose 10 per cent weight during accelerated thermal ageing and the rule of thumb concerning the change in rate of chemical reactions with temperature clearly indicate that the useful lifetime of plasticized PVC objects may be extended more than 10 times under ambient conditions by changing the storage environment from a polyethylene bag to a closed glass container or by low temperature storage (Table 7.1). Basing lifetime calculations on the time taken to lose 10 per cent of original weight has been one of the standards used by the plastics industry in relation to plasticized PVC film for many years (Quackenbos, 1954). Enclosing plasticized PVC objects, whatever their level of degradation, is inexpensive

Table 7.1 Time for plasticized PVC containing 25–30 per cent DEHP to lose 10% weight at 20°C calculated from weight loss measured during accelerated ageing (70°C)	
Environment	**Lightly plasticized PVC (days)**
enclosed in LDPE bag	256
open (not enclosed)	256
enclosed with activated carbon	320
enclosed with silica gel	576
enclosed with Ageless® oxygen absorber	2112
freezer enclosed in glass high relative humidity	>4160

to implement, of low practical complexity and allows public accessibility to plastics objects. Conservators and designers are advised either to improve ventilation or to include adsorbent materials to remove volatile degradation products from the air space surrounding plastics objects during storage. In the case of plasticized PVC, such action would accelerate the loss of plasticizer and thereby reduce the longevity of both new and deteriorated PVC objects.

7.1.3 Inhibition of degradation by low temperature storage

Storing plastics-containing objects at low temperatures (below 10°C) has been proposed as a relatively low cost technique for slowing the rate of the most common chemical degradation reactions, namely hydrolysis and oxidation (Michalski 2002). The justification for cold storage is that reducing the storage temperature by between 5–10°C halves the rate of chemical reactions. Some physical degradation processes are inhibited by cold storage. Reducing the storage temperature from ambient to that of a domestic freezer slows the rate of diffusion of plasticizer from polyvinyl chloride (PVC) by a factor of 15 (Shashoua, 2004).

At present, low temperature storage is only routinely applied to photographic archives and not to three-dimensional objects, so we have limited practical experience of its effects in real time. However, initial findings of a pilot study on the effect of cold storage on physical properties of selected plastics suggest that storage of plastics in a domestic freezer should be considered as an alternative to the present conservation options for storage (Shashoua, 2005).

Plastics materials exhibit both reversible changes (including dimensional and tensile) and irreversible changes (including extent of polymer crystallization and mechanical failure) on cooling. They contract or shrink considerably more than other materials found in museum collections such as metals, ceramics and glass. The linear coefficient of expansion for thermoplastics ($4.0–20.0 \times 10^{-5}\,°\mathrm{C}^{-1}$) is 5–10 times greater than those of most metals. For example, a copper pipe will shrink by 0.01 per cent if the temperature is reduced by 10°C. Under the same conditions, a high density polyethylene pipe shrinks by 0.07 per cent, and polypropylene and unplasticized polyvinyl chloride (PVC) pipes by 0.04 per cent (Nason et al., 1951). Although shrinkage of plastics is an unavoidable process on cooling, it is reversible and, in the absence of degradation, plastics will assume their original dimensions on return to ambient temperature.

The effect of shrinkage on cooling has greater significance when considering the physical stability of composites, that is objects which are constructed from several materials in close contact. As the composite is cooled, each material will attempt to shrink independently, but may be restrained by the others in contact. The magnitude of the tensile stress experienced by each material is a function of both the attempted shrinkage and the change in its tensile properties. The dimensional stability of photographic films, which are prepared from laminates of gelatine emulsion (containing information about the image) and plastic support layers, such as cellulose nitrate or acetate, has been extensively studied (Mecklenburg et al., 1994;Calhoun and Leister, 1959).

In addition to the effect of temperature on dimensional changes, the influence of the accompanying change in moisture content of the air surrounding the material must be considered. This is potentially a risk during the cooling process and while returning objects to room temperature after cold storage for materials which contain moisture. Research into cooling rolls of cellulose, acetate-based movie film concluded that a maximum difference of 10 degrees between film and its storage container or between any two areas in the mass of materials should be maintained to avoid formation of condensation (Padfield, 2002).

It is important to limit contact between liquid water (including that from condensation) and plastics, because some plastics have the ability to absorb moisture, swell and fail. Water acts as a plasticizer for many of the early plastics, notably casein, and can displace plasticizer from PVC (Shashoua, 2001). Polyamides, which include nylon, are the most hygroscopic polymers in common use, containing up to 3 per cent moisture by weight under ambient conditions. Cellulose acetate and poly (methyl methacrylate) contain 0.8 per cent, PVC contains 0.4 per cent and polystyrene 0.1 per cent moisture by weight at ambient.

The risk of introducing either irreversible physical changes or damage due to the formation of condensation into degraded plastics materials during the cooling process is not significant for plastics thinner than 1 cm (thin-walled), regardless of their polymer type or extent of degradation (Shashoua, 2005). Because polymers have a very low thermal conductivity, cooling of solid objects proceeds slowly. Thicker moulded plastics will require a longer period to attain the same temperature as a freezer in which they are placed than films, tubes or foams. For example, CN negative film, polyester cassette tapes and ABS Lego bricks attain freezer temperatures within 15–45 minutes. As a result, thin-walled plastics don't support the formation of condensation. In contrast, plastics which contain a bulk of material thicker than 1 cm (thick-walled) require at least 90 minutes for all areas to attain freezer temperature. The slower cooling period results in temperature differences between peripheral and central areas greater than Padfield's '10 degree difference' for the formation of condensation.

Methylene blue powder applied to filter paper for convenient handling is highly sensitive to the presence of liquid water, changing colour from grey to an intense blue colour (Figure 7.6). Condensation has been detected between the front cover and cockled first page of a PVC photograph album on cooling. The album is a thick-walled material and exhibited a temperature difference between centre pages and cover of around 30°C. Insulating plastics prior to placing in a freezer prolongs the cooling period, increases the maximum temperature difference within the materials and with it the possible formation of condensation.

Figure 7.6
Methylene blue indicator sheets can be used to detect the formation of condensation on cooling. Dark blue spots on the indicator sheet on the record itself suggest initial formation of condensation while the sheet on the cover remains grey.

Box 7.1   Glass Transition Temperature

Glass transition temperature (Tg) is a property of amorphous (low crystallinity) polymers whose molecules are not ordered in a regular way. Amorphous polymers include poly (methyl methacrylate), atactic polystyrene (methyl groups in the styrene monomer are arranged randomly along the polymer chain resulting in an irregular structure), polycarbonate, polyisoprene and polybutadiene. In order for molecules or segments of molecules to move, imparting bulk flexibility, there must be some free spaces in the material into which these molecules can move, simultaneously leaving spaces for other molecules to occupy. One interpretation of Tg is that it is a temperature below which the free space is too small for such molecular movement. Above the Tg, there is sufficient energy for molecular movement, jostling occurs and the free volume increases, resulting in polymers which are soft and flexible. When polymers are cooled below this temperature, they become hard and brittle, due to a great reduction in molecular mobility, and are in a glassy state.

Most plastics in collections have Tg values above ambient, indicating that they are in a glassy, stiff state at room temperature (Table 7.2). The exceptions are polyethylene, polypropylene, synthetic rubbers and plasticized PVC. On cooling from ambient to freezer temperature ($-20°C$), these materials undergo a change in state from rubbery to glassy. In their new state, their response to stress as a result of handling and moving changes, allowing the possibility to 'freeze in' molecular orientations, not present under ambient conditions. If the orientations vary from place to place in the structure, sufficient stresses may be established to cause the plastics material to distort. Polyurethanes are vulnerable to this phenomenon between $20°C$ and $-40°C$, indicating that handling them at low temperatures could cause irreversible damage (Byrne and Zukas, 1989). A PVC photograph album underwent a change in phase from rubbery at ambient temperature to glassy at freezer temperature, which caused stiffening of a photograph album's already crinkled pages (Figure 7.7).

Table 7.2 Glass transition temperatures (Tg) of plastics	
Plastic	**Tg (°C)**
polyethylene*	−20
polypropylene*	+5
poly (vinyl chloride) 20% plasticizer*	+20
poly (methyl methacrylate)	+45 to 115
poly (vinyl chloride) unplasticized	+80
polystyrene	+80 to 104
polytetrafluoroethylene	+115
*rubbery state in normal museum environment	

Although the phenomenon was reversed on warming to ambient, the distortion on cooling would have caused irreversible damage to any enclosed photographs or negatives. In my experience very degraded CN negative film loses flexibility and often breaks despite careful handling at freezer temperatures, so should not be manipulated while cold.

Figure 7.7

Plasticized PVC photograph album at 20°C (top) and at −20°C (middle). Pages of the album were cockled and stiffened on cooling (bottom). The distortion was reversed on return to ambient.

In conclusion, thin-walled cellulose nitrate, polystyrene, polyesters and acrylonitrile-butadiene-styrene copolymers may be safely stored in a freezer protected only by a closed polyethylene bag, despite their degraded conditions. Use of insulation to slow the rate of cooling offers no advantages for storage of plastics. Manipulation at freezer temperatures should be limited, particularly for degraded CN film and plasticized PVC, which become brittle when cooled. In addition, high shrinkage and risk of formation of condensation associated with degraded, plasticized PVC and composite materials provide less certainty as to the suitability of low temperature storage for these materials.

7.2 Interventive conservation

Interventive conservation treatments comprise invasive treatments which are applied as necessary to stabilize and strengthen degraded materials, to limit further degradation and often to preserve their original significance and value. Interventive treatments for plastics include cleaning, joining broken and failed components, consolidation, impregnation and filling. There is little in the conservation literature detailing interventive treatments for plastics and this subject is poorly developed compared with inhibitive conservation of these materials. This may be due to the high sensitivity of many plastics, especially when deteriorated, to organic liquids, aqueous solutions and water itself, which give interventive treatments the reputation of high risk and irreversible practices. Nevertheless it is often necessary to employ such treatments. This section discusses details of established interventive treatments from conservation and restoration literature and presents tools used by the plastics industry and others. Although applying registration numbers is not usually considered as conservation, it is usually an invasive procedure and may affect long-term stability in the case of plastics, so is discussed here.

7.2.1 Cleaning of plastics

Although surveys of the condition of museum collections containing plastics in the United Kingdom and Scandinavia have concluded that approximately 75 per cent of collections require cleaning, the practice is still poorly developed, largely due to the fear of damaging plastics irreversibly (Shashoua and Ward, 1995). Dirt on plastics usually comprises oily materials from handling and use prior to collection and particulates from air (Figure 7.8). Because plastics are good electrical insulators, they hold dust via static electricity. In addition, dirt may be present at surfaces as a result of degradation. Phthalate plasticizers migrate to surfaces of PVC with time and, in acidic or alkaline conditions, hydrolyze slowly to produce crystals of phthalic acid or anhydride (Wilson, 1995). If not removed, these can contaminate other objects or packing materials. In order to avoid build up of such contaminants, it is sensible to clean plastics regularly. A maximum cleaning interval of 5 years has been proposed for objects stored in stores without air filters (Morgan, 1993). Ideally, plastics in storage would never need cleaning more than once.

Figure 7.8
The polyurethane foam model of a shark used in the Jaws films became dirty with both use and degradation and appears discoloured as a result (upper image). Mechanical cleaning of selected areas improved its appearance (lower image).

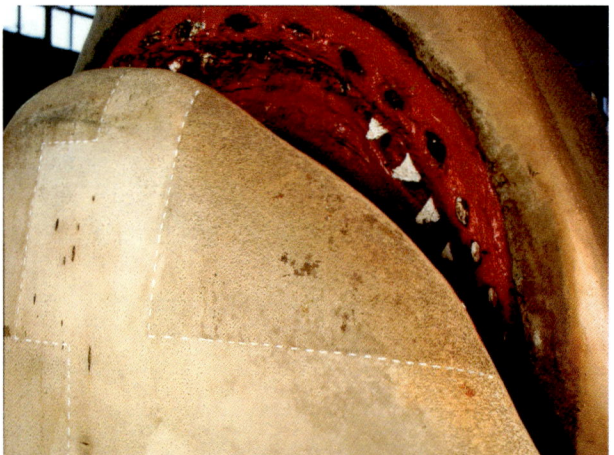

Cleaning techniques for all materials including plastics may be divided into mechanical and chemical. Chemical cleaning may be based either on aqueous or non-aqueous cleaning agents. The purpose of mechanical cleaning is to physically remove dust and residues from surfaces, either by blowing air from a can or cylinder, via suction from a vacuum cleaner, using artist's brushes or electrostatic cloths. Plastics exhibit poor resistance to abrasion. Those with Tg close to ambient, particularly plasticized PVC and polyethylene, are most likely to experience damage from mechanical cleaning processes. Scratches are areas with higher reactivity than their surroundings and are therefore more vulnerable to chain scission, reduction in tensile strength and permanent damage. In addition to a loss of physical properties, the presence of scratches changes the light reflecting properties of surfaces, resulting in a loss of gloss.

The purpose of chemical cleaning is to dissolve dirt, residues or other unwanted material at surfaces and displace it. The decision whether to use aqueous or non-aqueous cleaning agents depends both on the type of dirt present ('like dissolves like') and on the chemical properties of the plastic's surfaces. The type and polarity of the plastic, its condition and the presence of previous conservation treatments will influence the effectiveness of cleaning agents. For example, since hydrolytic breakdown is a major pathway for semi-synthetic polymers, the use of aqueous cleaning agents is generally inadvisable. Synthetic plastics are less likely to be deteriorated by a short period of contact with water. Manufacturers of slide rules made from filled cellulose nitrate dating from the 1930s and 1940s recommend cleaning them by brief immersion in a detergent solution, abrading with a domestic scourer and drying on filter paper or tissue because of water's efficiency at removing dirt from handling and use (Sphere Research Corporation, 2003).

In general, solvents are not recommended for cleaning plastics (Morgan, 1993). The risk from solvents swelling, dissolving and extracting additives from plastics is higher than from aqueous washing agents (Sale, 1995). Another phenomenon to be considered when cleaning rigid plastics such as polystyrene, polycarbonate and polymethyl methacrylate with solvents is stress cracking or crazing, correctly known as environmental stress cracking (ESC), which results in the development of white crystalline structures or interconnecting cracks in the body of the material (Fenn, 1993). Stress cracking is irreversible and may occur either immediately after application of solvent or may develop gradually after weeks or months. Chemicals that do not usually attack a polymer in an unstressed state can attack an area weakened by localized stress, causing a crack or craze. Such damage forms a fracture at micro scale and breaking of the polymer chains.

Polymer separation allows air into clear polymers. Due to differences between the refractive indices of air and polymers, the fractures are observed as whitening, crazing and cracks (Figure 7.9). The phenomenon is complex with several possible causes. One possibility is that absorption of high concentrations of solvent vapours induces compression and the release of strains introduced into the plastic during manufacture. Another option is that evaporation of the solvent causes heat loss to the plastic. Plastics have low thermal conductivities so while the cooled areas contract, adjacent areas retain their original temperature causing uneven changes in dimension.

Figure 7.9

Environmental stress cracking (ESC) can be induced by cleaning a rigid plastic with a low boiling point solvent. Acetone has been applied to the polystyrene jewel case shown here, resulting in the formation of opaque areas.

There are some occasions where it is necessary to apply solvents to plastics, for example removal of residues from adhesive labels or earlier repairs. Selection of liquids to clean plastics' surfaces, which will not soften or dissolve them, may be achieved in two ways. The trial and error approach involves applying drops of cleaning agents to a test piece or a hidden area of the object and examining the results with time. A more systematic approach is to compare the polarity of the liquid with that of the plastic in order to avoid those liquids which are most likely to dissolve or soften the object.

The term 'solubility parameter' was first used by Hildebrand and Scott in 1950 and describes systems used to quantify the principle of 'like dissolves like' (Hildebrand and Scott, 1950). Solubility parameters are experimentally determinable measures of the forces of attraction which hold molecules together. The Hildebrand solubility parameters have been further developed and refined by others, notably by Hansen, and applied to selection of additives for polymers, surface coatings formulation and a wide range of organic and inorganic materials (Hansen, 2007).

Solubility parameters provide a starting point for selecting cleaning agents (Table 7.3). A chemical will be a solvent for another if the molecules of the two materials prefer to be close together than to separate. Solution occurs when solubility parameters of polymer and solvent are within around $2\,MPa^{1/2}$. Based on solubility alone, it is clear that water is unlikely to damage the surfaces of plastics if used as a cleaning agent and that acetone-based materials are likely to soften or dissolve cellulose nitrate, poly (methyl methacrylate),

Table 7.3   Hildebrand solubility parameters for organic liquids and polymers. Solution occurs when solubility parameters of polymer and solvent are within around 2 MPa$^{1/2}$ of each other

Polymers	Hildebrand solubility parameter (MPa$^{1/2}$)
polytetrafluoroethylene	12.6
polyethylene	16.3
polypropylene	16.3
polybutadiene	17.1
polystyrene	18.7
poly (methyl methacrylate)	18.7
poly (vinyl chloride)	19.4
cellulose nitrate	21.6
poly (ethylene terepthalate)	21.8
cellulose acetate	23.2
nylon 66	27.8
Liquids	
hexane	14.9
turpentine	16.5
cyclohexane	16.7
carbon tetrachloride	17.6
xylene	18.0
toluene	18.2
ethyl acetate	18.6
chloroform	19.0
trichloroethylene	19.2
cellosolve	20.2
acetone	20.4
butanol	23.2
ethanol	26.0
glycerol	33.6
water	47.7

poly (vinyl chloride) and poly (ethylene terepthalate) so should be avoided. Solubility parameters for polymers change on degradation so published values for new materials provide only a guide.

In addition to selection of cleaning agents, selection of the technique by which they are applied depends on the properties of the surfaces to be cleaned. The use of moistened cotton swabs enclosed in lens tissue has been shown to minimize abrasion during aqueous cleaning of soft rubbers (Sale, 1995). Trials to clean a sculpture constructed from polyurethane ether (PURether) foam – *Funburn* made around 1969 by John Chamberlain and exhibiting discolouration, dirt, reduction in elasticity and loss of edges – indicated that a combination of mechanical and aqueous cleaning would be the most effective approach (Winkelmeyer, 2002). First, loose surface dirt including hair and dust particles were removed by brush and vacuum cleaner. Next, surfaces of the foam were wetted with deionized water applied by pump dispenser. A foam comprising non-ionic detergent (an alkylphenol polyglycol ether) at a concentration of 3 per cent by volume in deionized water was then applied with a paint roller. After 1 minute, the foam was absorbed with blotting paper, before rinsing with a deionized water spray and blotting dry.

Non-ionic detergent applied using cotton swabs at a concentration of 5 per cent has proved effective at removing superficial dirt from plasticized PVC (Huys and van Oosten, 2005). Surfaces of new and deteriorated plasticized PVC have been examined using non-contact laser surface profilometry (LSP) and contact angle measurements to determine changes in surface topography after cleaning with water and surfactant solutions. The study concluded that aqueous solutions did not introduce damage or significant levels of intrinsic change to PVC (Fairbrass, 1999).

In addition to removal of dust distributed evenly over surfaces, cleaning is also necessary to remove stains, spots, residues or scratches, whether they are degradation products or have an external cause. Collectors are particularly interested in removing stains because they reduce both commercial value and interpretation of artworks. Although the treatments suggested by plastic doll collectors are not established conservation techniques, they are practiced and so merit mentioning here. 'Green ear' affects vintage Barbie® and Sasha® dolls with metal earrings and corroded snap fasteners. The term refers to a greenish stain which is first seen on the PVC skin in the vicinity of dolls' earrings and spreads outwards onto the face or body (Figure 7.10). The cause is thought to be green corrosion products from copper-containing earrings or fasteners which is transported by plasticizer by the doll's PVC skin.

Some collectors apply commercial acne treatments daily to dolls for around six weeks to remove 'green ear'. Products such as Clearasil® (salicylic acid) and Oxy10® (benzoyl peroxide) are applied to the green stains and the dolls' heads are then placed in direct sunlight for 8–12 hours to accelerate the treatment.

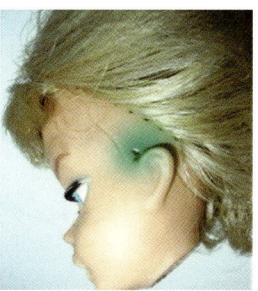

Figure 7.10

Collectors use commercial acne treatments to remove 'green ear' from dolls' PVC skin. Green ear comprises copper corrosion products formed on earrings, pins or metal fasteners, which are transported by plasticizer.

Salicylic acid forms a complex with copper, which is soluble in water so can be washed from the doll. The action of Oxy10® on 'green ear' is not documented, but benzoyl peroxide oxidizes copper ions so is likely to act as a bleaching agent and lighten the colour of the stain (Aleksandrov and Denisov, 1969). Dolls are rinsed in distilled water to stop the reaction (Locker, 1999). Collectors are advised not to apply more than one acne product to a doll. Because ultraviolet light is a degradation factor for PVC, its application may cause photodegradation, which is manifested by darkening.

Commercial polishes which claim to remove scratches from plastics are based either on abrasive particles suspended in a solvent or on a silicone polymer. The first physically smooths the edges of scratches in acrylic, polycarbonate and other rigid plastics when applied at right angles to the scratch. The second type of polish fills the scratch with a silicone polymer which wets the surfaces of flexible plastics, evens faults in surfaces and imparts high gloss. Because silicone polymers have very low surface tensions, it is very difficult to re-treat or remove polishes, so their use must be considered irreversible. Commercial polishes have not been evaluated for use on plastics in museum collections.

7.2.2 Joining plastics

The principle of irreversibility is difficult to apply to the process of joining plastics. It is almost impossible to successfully repair plastics without partial dissolution, melting or some other type of disruption to surfaces (Coxon, 1993). Joining of industrial plastics may be achieved by mechanical fastening, adhesive bonding, thermal- or solvent-welding. Very little has been published in conservation literature concerning evaluation or development of joining techniques. In the near absence of established conservation techniques, an industrial approach to joining plastics is presented here with the aim of providing a starting point for museum professionals working in this area.

Mechanical fastening of plastic components is usually achieved at the design stage and reserved for strong, flexible plastics which can withstand the strain associated with inserting screws, push-on clips and nuts. Thin sheets or films of flexible plastics can be joined by stitching in the same way as textiles, using synthetic thread such as polyester to maintain the change in dimension associated with changes in relative humidity and temperature.

The principles which are most relevant in describing adhesion of solid plastics are the mechanical and adsorption theories. Mechanical theory describes the surface profile of a plastic as a maze of microscopic peaks and troughs. In order to adhere effectively, the adhesive must penetrate the troughs and displace any trapped air or contaminants. Opportunities for interaction between adhesive and plastics surfaces can be improved by preparing surfaces mechanically to increase cleanliness, reactivity and surface area. The adsorption theory describes adhesion as a process resulting from contact at a molecular level between two materials and involves surface forces which develop between atoms in the two materials. For adhesion to occur, the adhesive must first wet the plastic adherend and spread over its surfaces (Figure 7.11). Good wetting results in a greater contact area between adherend and adhesive, over which the forces of adhesion may act, than poor or no wetting.

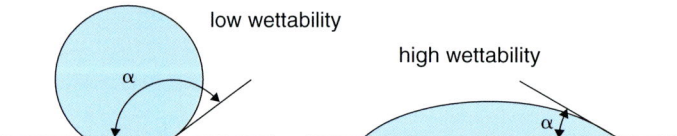

Figure 7.11

Left: poorly wetting liquids do not spread on the plastic's surfaces and are unlikely to result in adhesion. The contact angle (α) is greater than 90°. Right: good wetting is exhibited by liquids which make contact with surfaces over a large area and form a contact angle less than 90°.

Surface tension is defined as the force per unit length exerted by one surface. For an adhesive to adequately wet a plastic, it should have a surface tension lower than the plastic's surface tension. After intermolecular contact is achieved through good wetting, adhesion is primarily the result of formation of electrostatic forces including Van der Waals forces and hydrogen bonds. Since these bonds are approximately 10 times weaker than covalent bonds, they are only effective for a very short distance around 2–5 molecular diameters. Contact between adhesives and plastics surfaces to be joined must therefore be excellent.

Plastics have considerably lower surface tension values than metals. As a result, all polymeric adhesives can readily wet metal surfaces (Table 7.4). It is

Table 7.4 Surface tension values for common plastics and metals	
Material	**Surface tension mN/m at 20°C**
acrylics	32
acrylonitrile-butadiene-styrene	35
aluminium	c.500
cellulose (paper)	45
copper	c.1000
cyanoacrylates (superglues)	37
epoxy	47
polyamide	46
polycarbonate	46
polyethylene	31
poly (ethylene terephthalate)	43
poly (methylmethacrylate)	39
polystyrene	33
polytetrafluoroethylene	18
poly (vinyl chloride)	39
silicone	24
water	73

more difficult to find adhesives with lower surface tensions than plastic substrates, particularly those comprising polyethylene and polytetrafluoroethylene. A dilute, water-based adhesive is unlikely to wet plastic because the surface tension of water is 73 mN/m compared with only 30–45 mN/m for most plastics. A surface contaminated with silicone (18 mN/m) from, for example, previous treatment with waterproofing polish, will also be difficult to wet and subsequently to adhere. Plastics must be free of contaminants if adhesion is to be effective.

To achieve good wetting, some plastics require modifying by chemical or physical means to increase their surface tensions or energies. Polyethylene, polypropylene, polyesters and polytetrafluoroethylene are those plastics which most often require modifying. Correct selection of modification type is important (Table 7.5). For example, mechanical abrasion of polyethylene will roughen surfaces and introduce air pockets at adhesive-adherend interfaces, reducing contact between the two materials further. By contrast, oxidizing

Table 7.5	Pretreatments and adhesive types used by the plastics industry to join plastic components	
Plastic adherend	**Pretreatment/preparation**	**Adhesive types**
cellulose nitrate and acetates	clean with isopropanol	epoxies, polyurethanes, acrylics, cyanoacrylates (may cause ESC before curing)
phenol, melamine and urea formaldehydes	abrasion and solvent cleaning	epoxies acrylics, polyurethanes
plasticized poly (vinyl chloride)	clean with ketone	epoxies, polyurethanes, acrylics, cyanoacrylates
polyethylene and polypropyene	flame or plasma treatment to oxidise surfaces	epoxies
poly (methyl methacrylate)	clean mechanically	epoxies, cyanoacrylates, acrylics (higher tensile strength than substrate)
poly (ethylene terephthalate)	abrasion or solvent cleaning with toluene	epoxies, polyurethanes, polyesters
polystyrene	clean mechanically	water-based, solvent-free (particularly for polystyrene foams) or hot-melt
acrylonitrile-butadiene-styrene	clean mechanically	epoxies, acrylics, cyanoacrylate (higher tensile strength than substrate)
polycarbonate	clean mechanically	epoxies, urethanes and cyanoacrylates (higher tensile strength than substrate)
polyamide (nylon)	dry to less than 0.5% moisture content	epoxy (adhesion of nylon is unreliable)
polytetrafluoroethylene	plasma treatment	epoxy, polyurethanes (adhesion of PTFE is unreliable)

polyethylene by heating rapidly with hot air or an oxyacetylene flame will improve adhesion by toughening the material, thus increasing surface tension (Harper, 2000). Care is required when applying treatments involving heating, since polyethylene deforms and softens at around 100°C. Plasma treatment exposes plastics to an electrically activated inert gas producing a hard, crosslinked surface which is suitable for adhering. Plasma treatments are effective but costly.

Adhesion to plasticized PVC is often poorly effective because migrating plasticizer contaminates surfaces. In this case, pretreatment comprises cleaning with a ketone solvent such as acetone. Polystyrene, nylon, melamine-formaldehydes and polyesters can be pretreated by abrading with an emery cloth prior to cleaning with an alcohol. The effectiveness of all surface treatments decrease rapidly with time, so it is necessary to both modify and adhere within a short period.

Selection of an appropriate adhesive for joining industrial plastics is initially dependant on whether a structural material (high load-bearing properties, stable to heat, radiation, weathering and solvents) or a non-structural material (suitable to hold lightweight components in place) is required. If a structural adhesive is required, thermosetting adhesives such as epoxies, polyurethanes and acrylics are those most frequently chosen because they exhibit high toughness and flexibility (Table 7.5). Non-structurals include pressure-sensitive, contact and hot-melt adhesives. Secondary factors to consider when selecting adhesives for plastics are their thermal coefficients of expansion and changes in physical and chemical properties with time. When conserving outdoor artworks or objects stored in poorly controlled microclimates, large differences in thermal coefficient of expansion between the plastic adherend and adhesive can introduce stress at the joint interfaces. Physical or chemical degradation introduces changes at surfaces, which reduce the strength of the adhesive bond or displace the adhesive and cause bond failure. For example, oily phthalate plasticizers tend to migrate to PVC surfaces with time and can displace adhesives which are less compatible with the contaminated surfaces than with the clean ones.

The risk of loss of adhesion due to degradation of plastic substrates can be minimized by selecting a tolerant adhesive. Extensive evaluation of suitable adhesives to rejoin large sections of plasticized PVC used to construct the *Aeromodeller 00 PL*, a zeppelin balloon 28 m long and 6.5 m wide made by the Belgian artist Panamarenko in 1965, led to the use of an adhesive tape for the repair of small holes and tears and a liquid adhesive for larger sections. Tape comprising a PVC/ethylene vinyl acetate substrate backed with a styrene butadiene rubber adhesive was pressed into place, adhered well to the deteriorated PVC and proved resistant to yellowing. Rubber-based adhesive produced sufficiently flexible bonds with high peel strength to re-adhere 300 m of loose seams in the balloon (Huys and van Oosten, 2005).

Additional considerations for selecting an adhesive for plastics in collections are that:

- its refractive index (RI) is similar (within ± 0.01) to that of the adherend, particularly if the latter is transparent (Table 7.6)
- any thinning solvents do not induce ESC and their degradation pathways do not initiate or accelerate degradation of the substrate.

Table 7.6   Refractive indices (RI) of plastics and commonly used adhesives. A join is invisible if the adhesive matches the RI of the object ± 0.01. A consolidant or coating should match the RI of a surface ± 0.06 to give optimum saturation

Plastic type	Refractive index (n 20°C)
polytetrafluoroethylene	1.3500
polyvinyl acetate	1.4665
cyanoacrylate	1.4700
cellulose acetate	1.4757
poly (vinyl butyral)	1.4850
poly (methylmethacrylate)	1.4893
poly (vinyl alcohol)	1.5000
polyethylene	1.5100
cellulose nitrate	1.5100
plasticized poly (vinyl chloride)	1.5390
epoxy	1.5500
polyamide	1.5650
polycarbonate	1.5860
polystyrene	1.5894
phenol formaldehyde	1.7000

It is likely that the refractive index of plastics will change with degradation because most polymers yellow. The challenge of identifying an adhesive which was effective at adhering poly (methyl methacrylate), without causing ESC has been investigated extensively (Sale, 1993). Epoxy, ultraviolet curing acrylics and solvent-drying acrylics were found to be satisfactory.

Thermoplastics which do not degrade at melting point may be joined by applying heat or solvent to liquefy surfaces, a process known as heat-welding. The bond is formed via diffusion of liquid polymer between surfaces and produces a material which is generally 80–100 per cent as strong as the original plastic. The plastic component itself can either be used to create a bond, or a welding rod made of the same type of plastic as that to be joined can be melted at the interface and used to fill the gap. Heat may be applied via a soldering iron, hot plate, commercial heat sealer, gas-welding gun or flame. Required temperatures range from around 182°C for polyethylene to 343°C for polycarbonate (Harper, 2000).

Similar bond strength to that achieved by heat-welding can be achieved by solvent-welding or cementing. Suitable solvents have approximately the same solubility parameter as the plastic or plastics to be joined. Mixtures of solvents are frequently used to slow the rate of evaporation and thus avoid ESC crazing in plastics. Solvent is applied by brush or syringe and, when softened, the two parts are held together until the solvent has evaporated (Harper, 2000). Applying welding techniques to conservation practice requires consideration of risks of damage, irreversibility and change in the appearance and, perhaps, significance of the treated material.

7.2.3 Consolidating, impregnating and filling plastics

Polyurethane foams used in furniture upholstery often discolour and crumble, so have been the focus of several investigations to identify appropriate consolidants. Experience with both naturally aged and artificially aged polyurethane foams indicates that consolidation imparts future protection against degradation to the material, particularly if anti-ageing additives were included in the formulation (Winkelmeyer, 2002). Sturgeon glue, gelatine, methyl celluloses, acrylic dispersions and polyurethane dispersed in water have been evaluated and compared as consolidants for polyurethane foams (van Oosten, 2005).

One of the dilemmas involved in applying coatings to a cellular structure such as a foam is whether to replace the air in the cells with a filler, apply a thin film to 'seal' the cells or to coat only the cell walls. The first two options usually result in a thick, inflexible glossy coating which changes the appearance of surfaces and the elasticity of the foam. One solution is to apply coatings at a concentration of 5 per cent weight via a nebulizing system driven by air pressure, which produces droplets with volumes of 1–10 µm. The droplets are finer than those formed by spray guns, which produce droplets up to 100 µm. Smaller particles penetrate deeper into the cells than larger ones, coating the walls and leaving the cells open, allowing the foam to retain its elasticity. Before and after consolidation, samples of polyurethane ester were aged in the dark at 90°C and exposed to RH cycles of 35–80 per cent over a 3-hour period to initiate hydrolytic breakdown. Samples of foams were exposed to radiation from a Xenon-Arc lamp to initiate oxidation. A polyurethane dispersed in water was an effective consolidant because it inhibited discolouration and crumbling on ageing. The addition of Vitamin E as an antioxidant further improves the resistance of polyurethane ether foams to oxidation (van Oosten, 2005).

As well as providing physical support to a degraded plastic, impregnation can also be used as a technique to introduce anti-ageing additives. Epoxidized soya bean oil (ESBO) is a viscous, yellow liquid used commercially as an antioxidant, plasticizer and acid receptor. If ESBO could be introduced into plastics which deteriorate by producing acid, it could limit the build up of acid thus delaying the onset of autocatalysis. A cellulose nitrate mirror back from the 1920s, approximately 3 cm thick, exhibited crazing and was used as an experimental sample. Its lower half was immersed in ESBO. After 2 years, the half exposed to air had continued to degrade while degradation of the half immersed in ESBO had not. The researcher proposed that ESBO had diffused into the cellulose nitrate and trapped the nitrous oxides produced so that autocatalysis could not take place (Williamson, 1992). It has also been proposed that, while immersed in ESBO, oxygen and water could not readily get access to further degrade the plastic.

Filling losses in the furniture-object *Pratone* by Gruppo Strum in 1966, which resembles oversized blades of grass 95 cm high, required a material with similar flexibility and elasticity. A suitable filler for the polyurethane ether artwork was required to form a smooth surface which could accept a coating. These requirements were met with a blend of foam crumbs mixed with a heat activated adhesive based on acrylic and ethylene vinyl acetate polymers (Bützer, 2002). Excess adhesive was removed with a blotter. By varying the size and amount of crumbs, the hardness of the resulting filler could be controlled. Less adhesive resulted in a softer filler. Filler was placed in the missing areas, warmed to activate it and shaped. The fill could be removed by warming the adhesive. An identical filling practice has been applied to *Still life of watermelons* constructed from polyurethane ether foam by Piero Gilardi in 1967. The work represents a carpet of grass, watermelons and leaves. An acrylic adhesive was mixed with original foam crumbs to fill missing areas in leaves (van Oosten, 1999).

Where cracks in foams are sufficiently large to weaken the structure of an artwork, a filler which provides reinforcement as well as improving appearance may be necessary to minimize future damage. A relief by Ferdinand Spindel, which comprised a foam sheet (1.5 cm thick) folded irregularly and nailed to a chipboard base, exhibited large cracks. Implants of new polyurethane ester foam with the same cell diameter as the original were cut to size, secured in place with a polyvinyl alcohol dispersion and painted red to match the original surfaces. The structural fill is reversible (van Oosten, 1999).

7.2.4 Labelling plastics

When an object is acquired by a museum and registered in its collections, a registration number is applied to identify it for security and research purposes. Although numbering is not usually the responsibility of conservators, it is an invasive procedure and may affect long-term stability in the case of plastics, so is discussed here. The traditional approach is to first apply a barrier layer of diluted acrylic lacquer or other coating, write the number in ink solvent-based marker pen or paint on the dried barrier layer and, sometimes, to protect the number with a second layer of coating. The function of the barrier layer is twofold. It reduces the chance of ink migrating into the body of the object, where it may stain, and it facilitates removal of the registration number should it require alteration in the future.

Selection of a suitable barrier material for plastics raises identical challenges to selection of an adhesive or cleaning product. Solvent in the lacquer must neither dissolve the plastic substrate nor induce solvent cracking, while the lacquer must wet the plastic sufficiently well to produce a cohesive film. Both rigid and flexible plastics can be damaged by use of inappropriate barrier materials, causing either ESC, swelling or distortion.

Allowing insufficient time for solvents in the top coat to evaporate before returning objects to their original storage area can result in damage. It may take up to one month for acrylic solutions to through-dry depending on the surrounding rate of air flow and temperature. Fenn has evaluated the stability of various barrier materials and binding media used to apply registration numbers to deteriorating cellulose nitrate (Fenn, 1993). Dispersions of acrylic polymers were deemed the most suitable because they had glass transition temperatures above ambient, low organic solvent contents and neutral pH. Second best was shellac dissolved in ethanol.

Some dyes and pigments accelerate the rate of degradation of polymers when in contact. Nigrosine, an azine dye used in fabric marker pens, accelerates the degradation of cellulose nitrate (Hercules Inc., 1979). Carbon black does not react with plastics, whereas white pigments are less stable. Zinc oxide, calcium carbonate and aluminium pigments become hygroscopic or discoloured and are unstable in contact with degrading plastics (Fenn, 1993). Rutile titanium oxide has been identified as the most stable white pigment for labelling plastics. In conclusion, damage caused by solvents in contact with plastics during registration can be minimized by omitting a barrier layer and applying the number directly to the object using an ink based on an acrylic dispersion medium with carbon black or rutile titanium white as pigments.

Alternatives to applying numbers directly to plastics objects have been investigated. Labels bearing the registration number may be tied to objects with holes or handles. However, labels can become separated or lost and are not visually pleasing when an object is to be displayed. Adhesive labels are insufficiently permanent and often leave a disfiguring adhesive residue (Figure 7.12). Soft 2B–4B pencils can be used to mark polyethylene and poly (vinyl chloride).

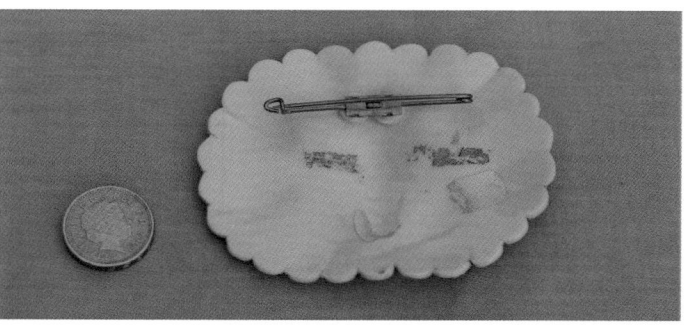

Figure 7.12

Adhesive paper labels often leave a disfiguring residue on plastics as shown on a polystyrene brooch from the 1960s.

Registration numbers may be mechanically etched into the plastic's surfaces and colouring materials applied afterwards to enhance visibility. The latter is an irreversible process and can be disfiguring.

References

Aleksandrov, A. L. and Denisov, E. T. (1969). Negative catalysis by copper ions in the chain reaction of cyclohexanol. *Russian Chemical Bulletin*, **18** (8), 1532–1536.

American Institute for Conservation of Artistic and Historic Works (AIC) (1997). *Commentary 20: Preventive Conservation* [online]. Available from: http://aic.stanford.edu/pubs/comment20.html [Accessed 2 April 2006].

Bradley, S. M. (1990). *A Guide to the Storage, Exhibition and Handling of Antiquities, Ethnographia and Pictorial Art*, 2nd edition. British Museum Publications, London.

Butler, B. L. (2006). *Cryovac® OS2000 Polymeric Oxygen Scavenging Systems* [online]. Available from: http://www.sealedair.com/library/articles/article-os2000.html. [Accessed 3 April 2006].

Bützer, J. (2002). Pratone – The resoration of a 1970s Polyurethane Flexible Foam Designer Seat. In *Postprints of the ICOM-CC Modern Materials Interim Meeting* Cologne, 12–14 March 2001 (T. van Oosten, Y. Shashoua and F. Waentig, eds) pp. 145–152, Siegls Fachbuch Handling.

Byrne, C. A. and Zukas, W. X. (1989). Shock-absorbing polymers: chemical analysis and characterization. *Polymeric materials science and engineering*, **61**, 560–564.

Calhoun, J. M. and Leister, D. A. (1959). Effect of gelatin layers on the dimensional stability of photographic materials. *Photographic Science and Engineering*, **3** (1). [online]. Available from: http://albumen.stanford.edu/library/c20/calhoun1959.html [Accessed 3 April 2006].

Conservation by Design (2007). Oxygen absorbers [online]. Available from: http://www.conservation-by-design.co.uk/oxyfree/oxyfree2.html [Accessed 10 August 2007].

Coxon, H. C. (1993). Practical pitfalls in the identification of plastics. In *Postprints of Saving the Twentieth Century: The Conservation of Modern Materials* Ottawa, 15–20 September 1991, (D. Grattan ed.) pp. 395–409, Canadian Conservation Institute.

Danish National Cultural Agency (Kulturarvstyrelsen) (2006)[online]. Available from: http://www.kulturarv.dk/forvaltning/museumsdrift/vejledninger/bevaring/index.jsp [Accessed 30 March 2006].

Davis, K. (2005). Pollution fighter turns clot buster. *New Scientist*, **5** February 2005, 25.

Derrick, M., Daniel, V. and Parker, A. (1994). Evaluation of storage and display conditions for cellulose nitrate objects. In *Preprints of the contributions to the IIC Ottawa Congress Preventive Conservation Practices, Theory and Research* Ottawa, 12–16 September 1994 (A. Roy and P. Smith, eds), pp. 207–211, IIC.

Derrick, M., Stulik, D. and Ornendez, E. (1993). Deterioration of cellulose nitrate sculptures made by Gabo and Pevsner. In *Postprints of Saving the Twentieth Century: The Conservation of Modern Materials* Ottawa, 15–20 September 1991 (D. Grattan, ed.) pp. 169–182, Canadian Conservation Institute.

Fairbrass, S. (1999). Surface degradation of poly (vinyl chloride). PhD thesis, Department of Chemical Engineering and Chemical Technology, Imperial College of Science, Technology and Medicine, London University.

Fenn, J. (1993). Labelling plastic artefacts. In Postprints of Saving the Twentieth Century: The Conservation of Modern Materials Ottawa, 15–20 September 1991 (D. Grattan, ed.), pp. 341–350, Canadian Conservation Institute.

Fenn, J. (1995). The cellulose nitrate time bomb: using sulphonephthalein indicators to evaluate storage strategies. In *Postprints Tate Gallery conference 'From Marble to Chocolate'. The Conservation of Modern Sculpture* London, 18–20 September 1995 (J. Heuman, ed.) pp. 87–92, Archetype Books.

Fuchs, D. R. and Leissner, J. (1995). Glassensoren erfassen das Schadensrisiko an Kunstobjekten. Restauro, **3**, 170–173.

Grattan, D. W. and Gilberg, M. (1994). Ageless® oxygen absorber: chemical and physical properties. *Studies in Conservation*, **39** (3), 210–216.

Hansen, C. M. (ed.) (2007). *Hansen Solubility Parameters: A user's handbook*, 2nd edition CRC Press, New York.

Harper, C. A. (2000). *Modern Plastics Handbook*. McGraw-Hill.

Hatchfield, P. B. (2002). Pollutants in the Museum Environment. Archetype Publications.

Hercules Inc. (1979). Nitrocellulose. *The Chemical and Physical Properties*, Technical pamphlet, Hercules Inc., Wilmington, Delaware.

Hildebrand, J. and Scott, R. L. (1950). *The Solubility of Nonelectrolytes*, 3rd edition. Reinhold, New York.

Huys, F. and van Oosten, T. B. (2005). The 'Aeromodeller 00-PL'; the conservation of a PVC balloon. In *Preprints of the 14th ICOM-CC Triennial Meeting* The Hague, 12–16 September 2005 (I. Verger, ed.) pp. 335–342. James&James Ltd.

ISO (2002). ISO 18925:2002: *Imaging materials – Optical disc media – Storage practices*. International Organization for Standardization.

ISO (2003). ISO 11799:2003: *Information and documentation – Document storage requirements for archive and library materials*. International Organization for Standardization.

Kessler, K. (2002). Testing environmental conditions for the storage of celluloid objects. In *Postprints of the ICOM-CC Modern Materials Interim Meeting* Cologne, 12–14 March 2001 (T. van Oosten, Y. Shashoua, and F. Waentig, eds) pp. 134–144, Siegls Fachbuch Handling.

Ligterink, L. J. (2002). Notes on the use of acid absorbents in storage of cellulose acetate-based materials. In *Contributions to Conservation, Research in Conservation at the Netherlands Institute for Cultural Heritage* (J. A. Mosk and N. H. Tennent, eds) pp. 64–73, James&James Ltd.

Locker, S. (1999). *Green Ear? The cause and treatment* [online]. Available from: http://collectdolls.about.com/library/weekly/aa050299.html [Accessed 30 March 2006].

Lovett, D. and Eastop, D. (2004). The degradation of polyester polyurethane: preliminary study of 1960s foam-laminated dresses. In *Preprints of IIC Congress Modern Art, New Museums* Bilbao, 13–17 September 2004 (A. Roy and P. Smith, eds) pp. 100–104, IIC.

Mecklenburg, M. F., McCormick-Goodhart, M., and Tumosa, C. S. (1994). Investigation into the degradation of paintings and photographs using computerized modelling of stress development. *Journal of the American Institute for Conservation* [online] **33** (2), 153–170. Available from: http://aic.stanford.edu/jaic/articles/jaic33-02-007_3.html [Accessed 2 April 2006].

Michalski, S. (2002). Double the life for each five-degree drop, more than double the life for each halving of relative humidity. In *Preprints of the 13th ICOM-CC Triennial Meeting* Rio de Janeiro, 22–27 September 2002 (R. Vontobel, ed.) pp. 66–72, James&James Ltd.

Morgan, J. (1993). A joint project on the conservation of plastics by The Conservation Unit and the Plastics Historical Society. In *Postprints of Saving the Twentieth Century: The Conservation of Modern Materials* Ottawa, 15–20 September 1991 (D. Grattan, ed.), pp. 43–50, Canadian Conservation Institute.

Nason, K., Carswell, T. S. and Adams, C. H. (1951). Low temperature behavior of plastics, *Modern Plastics, December.* 127–203.

Padfield, T. (2002). Condensation in film containers during cooling and warming. In *Postprints of Preserve, then Show Copenhagen*, October 2001 (D. Nissen, L. R. Larsen, T. Christensen and J. S. Johnsen eds) pp. 1–9, The Danish Film Institute.

Parmar, S. S. and Grosjean, D. (1991). Sorbent removal of air pollutants from museum and display cases. *Environmental International*, **17**, 39–50.

Quackenbos, H. M. (1954). Plasticizers in vinyl chloride resins. *Industrial and Engineering Chemistry*, **46**(6), 1335–1341.

Quye, A. (1999). Care Advice Summary. In *Plastics collecting and conserving* (A. Quye and C. J. Williamson, eds) pp. 136–137. NMS Publishing Ltd.

Ram, A. T., Kopperl, D. F. and Sehlin, R. C. (1994). The Effects and Prevention of Vinegar Syndrome. *The Journal of Imaging Science and Technology*, **38** (3), 249–261.

Sale, D. (1993). An evaluation of eleven adhesives for repairing poly(methyl methacrylate) objects and sculpture. In *Postprints of Saving the Twentieth Century: The Conservation of Modern Materials* Ottawa, 15–20 September 1991 (D. Grattan, ed.) pp. 325–339, Canadian Conservation Institute.

Sale, D. (1995). Standing out like a sore thumb: a damaged sculpture made of three synthetic polymers. In *Preprints of From Marble to Chocolate, the Conservation of Modern Sculpture* London, 18–20 September 1995 (J. Heumann, ed.) pp. 89–103, Archetype Publications.

Shashoua, Y. (1999). Degradation and conservation of plasticized poly (vinyl chloride) objects. In *Preprints of 12th ICOM-CC Triennial Meeting* Lyon, 29 August–3 September 1999 (J. Bridgland, ed.) pp. 881–887, James & James Ltd.

Shashoua, Y. R. (2001). Inhibiting the degradation of plasticized poly(vinyl chloride) – a museum perspective. Ph.D. thesis, Danish Polymer Centre, Technical University of Denmark.

Shashoua, Y. (2004). Modern Plastics–do they suffer from the cold. In *Preprints of IIC Congress Modern Art, New Museums* Bilbao, 13–17 September 2004 (A. Roy and P. Smith eds) pp. 91–95, IIC.

Shashoua, Y. (2005). Storing plastics in the cold – more harm than good?. In *Preprints of the 14th ICOM-CC Triennial Meeting* The Hague, 12–16 September 2005 (I. Verger, ed.) pp. 358–364, James & James Ltd.

Shashoua, Y. and Thomsen, S. (1993). A field trial for the use of Ageless in the preservation of rubber in museum collections. In *Postprints of Saving the Twentieth Century: The Conservation of Modern Materials* Ottawa, 15–20 September 1991 (D. Grattan, ed.) pp. 363–372, Canadian Conservation Institute.

Shashoua, Y. and Ward, C. (1995). Plastics: modern resins with ageing problems. In *Preprints of Resins, Ancient and Modern* Aberdeen, September 1995 (M. M. Wright and J. H. Townsend eds) pp. 33–37, SSCR.

Sphere Research Corporation (2003). *Cleaning and Caring for Slide Rules* [online]. Available from: http://sphere.bc.ca/test/clean.html [Accessed 31 March 2006].

van Oosten, T. B. (1999). Plastic Surgery:conservation treatments for flexible polyurethane foams: from face-lift to donating the corpse to science. In *Preprints of*

Reversibility – does it Exist London, 1 July 1999, British Museum Occasional paper no.135 (W. A. Oddy and S. Carroll, eds) pp. 33–36, British Museum Publishing.

van Oosten, T. B. (2005). Conservation of plastics. In *Preprints of Australian Institute for Conservation of Cultural Material Objects Special Interest Group (SIG) Symposium and Workshop* Melbourne, 23–24 August 2005 (A. Pagliarino, ed.) pp. 4–19, AICCM.

Ward, C. and Shashoua, Y. (1999). Interventive Conservation Treatments for Plastics and Rubber Artefacts in the British Museum. In *Preprints of 12th ICOM-CC Triennial Meeting* Lyon, 29 August–3 September 1999 (J. Bridgland ed.) pp. 888–893, James&James Ltd.

Williamson, C. J. (1992). 150 years of plastics degradation. In *Postprints Polymers in Conservation* Manchester, 17–19 July 1991 (N. S. Allen, M. Edge and C. V. Horie, eds) pp. 1–13, Royal Society of Chemistry.

Wilson, A. S. (1995). *Plasticisers: Principles and Practice.* The Institute of Materials.

Winkelmeyer, I. (2002). Perfection for an instant-restoration of a polyurethane soft sculpture by John Chamberlain. In *Postprints of the ICOM-CC Modern Materials Interim Meeting* Cologne, 12–14 March 2001 (T. van Oosten, Y. Shashoua, and F. Waentig, eds) pp. 153–164, Siegls Fachbuch Handling.

The future of plastics conservation

Summary

The starting point for Chapter 8 is the fact that plastics conservation is a new specialism which requires further development. Progress has been slow partly due to a shortage of formal training for conservators in this area. In addition, the challenges of adhering to traditional ethical guidelines of conservation, particularly the principle of reversibility of interventive treatments, has limited the opportunities for development. As the 20th anniversary of plastics conservation as a recognized discipline approaches, there is growing need for more established techniques. Areas for development include monitoring of deterioration during storage and display, more effective conservation treatments, greater opportunities to study plastics and better accessibility to technical data concerning plastics.

The study of deterioration and conservation of plastics in museums, galleries and private collections started in a structured way in the early 1990s (Grattan, 1993). Conservation of plastics is one of the newest disciplines of the conservation profession. Growth in the field can be quantified by a fourfold increase in membership of the International Council of Museums' Committee for Conservation's Modern Materials and Contemporary Art working group between 1999 and 2006. However, fewer than 20 museum professionals are full-time specialists in plastics in Europe. One consequence of the lack of expertise in relation to the growing number of deteriorating plastics is that progress in the field of plastics conservation is slow.

In addition to a deficit in the number of conservation professionals compared to the number of deteriorating plastics in collections, the development of new conservation techniques for plastics has been impeded by a lack of available information about real time degradation pathways of fully-synthetic plastics, by difficulties of applying traditional, ethical principles of conservation to new, short-lived materials and by insufficient professional training courses and information for those using and conserving plastics (Keneghan, 2005). Progress in these areas must be made if plastics conservation is to develop further.

8.1 Monitoring the condition of plastics in collections

Because condition surveys require a high investment in resources, they are not usually repeated at regular intervals and therefore provide only a snapshot of condition at one particular moment. Plastics have an induction period following manufacture, during which no physical or chemical changes can be detected, followed by a period during which irreversible damage develops and can be measured. The shift from induction to degradation periods is short. As a result, plastics which do not exhibit degradation and which are documented as requiring no conservation can be irreversibly damaged six months later. Regular condition surveys would provide sufficient data to map the rate of degradation time for a collection. Such maps could be used to develop conservation strategies for collections and enable conservation managers to plan for future needs.

Identification of degradation markers for each plastic could be used as early warning detection systems for materials in storage or on display. A degradation marker is a chemical compound or physical or chemical property which, when quantified, reflects the extent of degradation of a plastic. The quantity of marker must be proportional to the extent of degradation and not affected by usage of the plastic. As an example, the concentration of acetic acid is used as a degradation marker for cellulose acetate, particularly in movie film bases. As cellulose acetate hydrolyzes, increasing concentrations of acetic acid are produced. Acetic acid is produced before signs of deterioration are visible. Acetic acid (vinegar) can be detected by odour and by measuring pH. Acid-Detection (A-D) strips are commercially available, acid-based indicator papers that change colour in the presence of the acidic vapour given off by degrading cellulose acetate movie film. Comparison of the colour of the A-D strip with that of the accompanying reference scale provides an objective way to quantify the extent of degradation of cellulose acetate (Image Permanence Institute, 2007).

Potential degradation markers for video and audio tapes, which contain polyester and polyurethane, have been investigated recently as part of the Presto Space project (Thiebaut et al., 2006). Chemical markers investigated included pH of the tapes' surfaces, which was expected to fall with the onset of hydrolysis, while mechanical markers included the loss of cohesion between the various materials which comprise tapes. Because few cultural institutions and private collectors have access to analytical instruments, effective degradation markers should be based on readily detectable properties such as colour change or colour change induced in another material, and the evolution of volatiles which results in odour or change in acidity.

8.2   Research into the deterioration of synthetic plastics

Surveys of plastics in many national museum collections suggest the least stable plastics to be cellulose nitrate, cellulose acetate, plasticized poly (vinyl chloride) and polyurethane foam.

Because of lack of resources, these 'high-risk' materials have dominated conservation research projects at the exclusion of others. Inhibitive conservation techniques have been established for them (Shashoua, 2006). Polyester, polyethylene and polypropylene plastics are increasingly showing signs of degradation in museum collections and should be the focus of more studies than they are today (van Oosten et al., 2008). The latest polymer – polycarbonate – became commercially available at the start of the 1960s and no new industrial polymer types have been developed since. As a result, the number of new polymers entering collections is limited, although not the number of plastics formulations or objects.

The major factors which initiate and accelerate degradation pathways for all fully-synthetic polymers need to be identified if effective conservation treatments for them are to be developed. In addition, recycled plastics occupy increasing proportions of everyday plastics as well as art in the form of carrier bags, furniture, collage and sculpture. As a result, they are likely to be collected in increasing amounts. Recycled plastics may comprise a single material or a mixture of thermoplastics. The useful lifetimes of recycled plastics have not been established, but are likely to be shorter than new equivalents due to the high temperatures applied during recycling treatments.

Research into degradation pathways of plastics has focussed to date on the polymer component of plastics. The fate of additives with time, with the exception of phthalate plasticizers and a few antioxidants, has largely been neglected and would benefit from study. Some pigments and dyes, heat and light stabilizers and plasticizers have been replaced with less toxic versions since the 1970s and their stabilities and likely interaction with other components in plastics formulations should be investigated. An example of a plasticizer which has been proposed as an alternative to phthalates is synthesized from hardened castor oil and acetic acid (ElAmin, 2006). Manufacturer Danisco claim Grinsted Soft-N-Safe® to be colourless, odourless and biodegradable. The latter property suggests that its useful lifetime may be shorter than that of the phthalate it replaces.

8.3   Development of additional conservation techniques

Conservation of plastics by inhibiting degradation pathways is possible today but limited in scope and effectiveness. Slowing the rate of degradation of most plastic types is realized by storing objects in an appropriate microclimate which removes the main factors causing degradation such as oxygen, acidic gases and moisture. Broad-spectrum, poorly-specific adsorbents such as activated carbon and zeolites are currently used to achieve microclimates. They are inefficient and require frequent renewal, which is excessively resource demanding for many museums. Research is needed to identify more specific adsorbents such as those employed by the medical and petrochemical industries.

Another option for inhibiting deterioration is to store plastics at temperatures below ambient. At present, low temperature storage is only routinely applied to photographic archives and not to three-dimensional plastics objects, so we have limited practical experience of its effects in real time. Pilot studies on the effect of low temperature on the physical properties of selected plastics indicate that storage of plastics in a domestic freezer is an alternative to the present inhibitive conservation options for long term-storage (Shashoua, 2005).

Although condition surveys of plastics in museums suggest that approximately 75 per cent require cleaning, the practice is still poorly developed (Shashoua and Ward, 1995). All interventive conservation techniques for plastics are underdeveloped compared with inhibitive treatments, due to the high sensitivity of many plastics, especially when deteriorated, to cleaning agents, solvents, adhesives and consolidants. In addition, coatings or adhesives which adhere successfully to plastics' surfaces either soften the substrate by dissolving or etching them to bond. Such treatments frequently change the appearance of the original, which may not be acceptable for cultural property, particularly works of art. Commercial protective coatings based on nanotechnology are available to protect vehicles and textiles from stains without changing their appearance. These should be investigated to evaluate their ability to protect and inhibit degradation by weathering for outdoor plastics.

Research into interventive conservation techniques is likely to be most effective when conducted as multidisciplinary projects involving conservators, conservation scientists, historians or curators, industrial chemists, artists and designers. While conservation scientists, industrial chemists and designers have knowledge of properties of materials and conservators can contribute with the practical aspects of applying materials and techniques, historians and artists can advise on the ethical and moral rights aspects of any interventive treatments proposed. With time and experience, guidelines or a charter could be proposed to define the approach and limits to be considered when applying interventive conservation to contemporary art or plastics materials.

Pressure, particularly from the commercial art market, private collectors and organizers of exhibitions, is likely to result in more research into interventive conservation as well as restoration treatments. In the past, privately funded plastics research has targeted specific artworks or collections rather than plastics types (Albus et al., 2007). Where findings are made accessible to all interested in the conservation of plastics, both object-based and material-based investigations are equally useful.

8.4   Training in plastics degradation and conservation

Today, there are no full-time degree or post-graduate courses in Europe for conservators or conservation scientists wishing to specialize in plastics. However, there are modules in plastics deterioration and conservation, and student projects in plastics at degree and master level offered by several European institutes including Cologne University of Conservation Sciences in Germany, EVTEK Institute of Arts and Design in Finland (EVTEK, 2007) and the V&A/Royal College of Art (V&A/RCA, 2007). In addition, short professional development courses in conservation of plastics are available at several institutes including West Dean College in the UK and the Campbell Center for Historic Preservation Studies and the Northern States Conservation Center in the USA.

Education in plastics conservation was initiated in the 1990s and the number of courses is growing steadily. Specialized, full-time courses for plastics conservators at undergraduate level would ensure that museums and other cultural institutes had qualified staff to conserve and research their collections. A combination of formal training and more practical internships involving conservation of plastics, particularly in contemporary art, modern history and medical collections, would equip conservators and conservation scientists with knowledge and experience of new materials.

8.5   Development of an information interface between plastics manufacturers and users

Some artists require data from which to select commercial plastics with the desired chemical and physical properties, including longevity, and those with minimal health and safety risks. Artists may use commercial products for another application than that intended so it is important to know their properties. Conservation professionals and historians need technical data about the major causes of deterioration and deterioration pathways for plastics. Such information would help to estimate the useful lifetimes of objects containing plastics prior to acquisition and develop a conservation strategy for the collection. Conservators may also need to select filling materials or support materials for degraded plastics objects.

Companies selling plastics in Europe are obliged to supply technical or data sheets with their products detailing health and safety risks and selected chemical and physical properties. One can search electronically a database of material data sheets such as MatWeb to find plastics with the desired physical or chemical properties, polymer type or trade name (MatWeb, 2007). Data sheets usually describe properties, processes and chemical structures in specialist terms understandable to a polymer chemist. In addition, data sheets name only the major ingredients in plastics formulations and do not always define their function. Such information is rarely accessible or useful to artists and conservators.

Artists and conservators would benefit from a searchable, reliable source of material information presented in a way that was understandable to non-chemists. Such a database would serve as a materials advisory service. 'Plasticsnetwork.org' is a website which resulted from an international collaborative project to promote an understanding of plastics and design and was coordinated by the Arts Institute at Bournemouth, UK (plasticsnetwork.org, 2005). It includes information about the history of plastics, manufacture, general properties and design applications from plastics historians, designers, plastics manufacturers and the British Plastics Federation, the professional body for the plastics industry in the UK. It does not include details of physical and chemical properties of polymer types, so cannot be used to select specific commercial products material   s, but provides sufficient information to allow selection of plastics types. More of these type of databases are required if the plastics industry is to successfully communicate with artists and conservators.

Some art schools and universities in Finland and Denmark are starting to include courses in polymer chemistry and physics to help students select materials which have the required useful lifetime while minimizing health risks to the user. Conservator Louise Cone of the Danish National Gallery has suggested that conservators and conservation scientists could collaborate with artists who wish to select those plastics with long useful lifetimes and the greatest possibility to be conserved in future. Such collaboration would benefit artists as well as museums and galleries because it could result in higher quality materials being selected than is the case today.

References

Albus, S., Bonten, C., Keßler, K., Rossi, G. and Wessel, T. (2007). *Plastic Art – A Precarious Success Story*. AXA Art Versicherung, Cologne.

EVTEK Institute of Arts and Design (2007). *EVTEK Institute for Arts and Design* [online]. Available from: http://www.evtek.fi [Accessed 3 August 2007].

ElAmin, A. (2006). *Biodegradable plasticizer developed as phthalate replacement*. Available from: http://www.packwire.com/news/printNewsBis.asp?id=64903 [Accessed 7 March 2007].

Grattan, D. W. (ed.) (1993). *Saving the twentieth century: the conservation of modern materials*. Canadian Conservation Institute, Ottawa.

Image Permanence Institute. (2007). *A-D Strips* [online]. Available from: http://www. imagepermanenceinstitute.org/shtml_sub/cat_adstrips.asp [Accessed 2 August 2007].

Keneghan, B. (2005). Plastics preservation at the V&A. *V&A Conservation Journal* [online] No. 50, Summer 2005. Available from: http://www.vam.ac.uk/res_cons/conservation/journal/number_50/plastics/index.html [Accessed 2 July 2007].

MatWeb.com. (2007). *Material propert in MatWeb* [online]. Available from: http://www.matweb.com [Accessed 2 August 2007].

Plasticsnetwork.org. (2005). *Plasticsnetwork homepage* [online]. Available from: http//www.plasticsnetwork.org [Accessed 5 August 2007].

Shashoua, Y. (2005). Storing plastics in the cold – more harm than good? In *Preprints of the 14th ICOM-CC Triennial Meeting* The Hague, 12–16 September 2005 (I. Verger, ed.) pp. 358–364, James&James Scientific publishers. London.

Shashoua, Y. (2006). Plastics. In *Conservation Science Heritage Materials* (E. May and M. Jones, eds) pp. 185–210. RSC Publishing, Cambridge.

Shashoua, Y. and Ward, C. (1995). Plastics: modern resins with ageing problems. In *Preprints of Resins, Ancient and Modern Aberdeen*, September 1995 (M. M. Wright and J. H. Townsend, eds) pp. 33–37. SSCR.

Thiebaut, B., Vilmont, L.-B. and Lavedrine, B. (2006). *D6.1: Report on video and audio tape deterioration mechanisms and considerations about implementation of a collection condition assessment method (D6.2)* [online] http://www.prestospace.org/project/deliverables/D6-1.pdf [Accessed 2 August 2007].

van Oosten, T., Bollard, C., de Castro, C. (2008). Lights out!!! Research into the conservation of polypropylene. In: *Postprints of Plastics: Looking at the past & learning from the future* London, 23–25 May 2007, publication expected in 2008.

V&A Museum/Royal College of Art (2007). *MA in Conservation Post-Graduate Training and Research* [online]. Available from: http://www.vam.ac.uk/school_stdnts/stdnts_lecturers/rca_va_courses/index.html [Accessed 2 August 2007].

Appendix 1
Optical, physical, thermal and chemical properties of the most frequently collected plastics

Data in the following tables refer to unfilled, non-reinforced plastics. The major sources of data provided are Pedersen (1999) and Brydson (1999). Materials are arranged in the tables in alphabetic order.

Ignition time is not determined by a standard test and was determined using the following procedure. A standard block (1.66 mm × 1.66 mm × 1.66 mm) was heated electrically and the time taken for it to ignite was measured in seconds. The shorter the period, the higher the flammability of the plastic.

Name/Abbreviation	Cellulose acetate (CA)	
Repeat unit		
Trade names and synonyms	Tricel®, Celanese®, Kodacel®, Rayon®, Tenite®, Similoid®, safety film, acetate	
History	First synthesized in 1865. Fibres first produced commercially in 1919.	
Major applications	Available as film, moulded and extruded forms and as fibres and textiles.	
	Photographic film, packaging films, graphic tracing films, display boxes, spectacle frames, hair brushes, combs.	
Optical properties	High transparency and available in many colours.	
	Refractive index (ND 20°C)	1.46–1.50
	Light transmission (%)	88
Physical and thermal properties	CA polymer is rigid and requires plasticizers. Dimethyl phthalate, triacetin and triphenyl phosphate have been used commercially. CA is tough and it is thermoplastic. CA is not as effective as fully-synthetic polymers in electrical insulation.	
	Density (g/cm^3)	1.3
	Tg (°C)	120
	Tm (°C)	232
	Tensile strength at break (MPa)	15–65
	Elongation at break (%)	5–8
	Permeability to oxygen at 30°C (10^{10} cm^3 s^{-1} mm cm^{-2} cmHg^{-1})	7.8
	Permeability to water at 25°C (10^{10} cm^3 s^{-1} mm cm^{-2} cmHg^{-1})	75000
	Moisture absorption at 21°C/80%RH for 24h (%)	2.1–3.2
Chemical properties	Higher water absorption than fully-synthetic polymers.	
	Degrades in the presence of light and heat. Plasticizer is lost causing shrinkage and distortion. CA has poor chemical resistance and is attacked by acids and concentrated alkalis producing corrosive acetic acid, which promotes autocatalysis. Hydrolytic breakdown accelerated by metals, particularly copper. Soluble in ketones and chlorinated solvents.	
	Degree of crystallinity (%)	low

Name/Abbreviation	Cellulose nitrate (CN)	
Repeat unit		
Trade names and synonyms	Xylonite®, Parkesine®, Durofix®, Celluloid®, nitrocellulose, nitrocotton, gun cotton, pyroxylin, French ivory	
History	First synthesized in 1832–62 (depending on source). Manufacturing process patented in 1865.	
Major applications	Available in film, moulded and extruded forms, fibres and lacquers.	
	Up to the 1950s: photographic film, bicycle parts (mudguards, pump covers), shadow puppets, dolls, spectacle frames.	
	Today: table-tennis balls, knife handles, guitar picks, carrier in topical medicines, filters for biological materials, nail polish, vehicle repair coatings.	
Optical properties	Water-white transparency and excellent clarity due to low crystallinity.	
	Birefringent.	
	Refractive index (ND 20°C)	1.49–1.51
	Light Transmission (%)	88
Physical and thermal properties	Very stiff as raw polymer so plasticizer is required to increase flexibility. Camphor was added as plasticizer until the 1950s and was then replaced by phthalates. CN is thermoplastic. It is highly flammable. CN exhibits good rigidity and toughness. It can after-shrink around inserts, so is useful for tool handles. CN is highly polar, which gives it poor electrical insulation properties. It has high permeability to gases and vapours.	
	Density (g/cm³)	1.35–1.40
	Tg (°C)	145–152
	Tm (°C)	100 (softens)
	Ignition time(s)	5
	Tensile strength at break (MPa)	35–70
	Elongation at break (%)	10–40
	Coefficient of thermal expansion $(°C)^{-1} \times 10^{-6}$	119–360
	Moisture absorption at 21°C/80%RH for 24 h (%)	1.0
Chemical properties	Plasticizer is lost with time causing shrinkage and distortion. CN has poor chemical resistance and is attacked by acids and alkalis producing corrosive nitrogen oxides which promote autocatalysis. Hydrolysis is accelerated by metals especially copper and zinc. Soluble in polar solvents including esters, ketones alcohols and ethers. Discolours on exposure to UV light and heat. CN becomes brittle and disintegrates on thermal and light ageing.	
	Degree of crystallinity (%)	low

Name/Abbreviation	Epoxy
Repeat unit	
Trade names and synonyms	Araldite®, Epoxy BK®
History	The first commercial epoxies were prepared in 1927 in the USA by reacting epichlorohydrin with bisphenol A. The name epoxy is based on the Greek words *epi*, which means upon and *oxy*, which means sharp/acidic.
Major applications	Adhesives, surface coatings, encapsulating and moulding compounds when reinforced, boat, vehicle and aircraft parts, patterns and moulds for shaping thermoplastics, sports equipment.
Optical properties	Water-white
	Refractive index (ND 20°C)

Refractive index (ND 20°C)		1.47–1.50

Physical and thermal properties for Nylon 6,6 poly (hexamethylene adipamide)	Epoxies have excellent electrical and thermal resistance. They are thermosets. Addition of fillers improves hardness, impact resistance and thermal conductivity.	
	Density (g/cm^3)	1.1–1.4
	Tm (°C)	155–195
	Ignition time(s)	>300
	Tensile strength at break (MPa)	28–90
	Elongation at break (%)	3–6
	Coefficient of thermal expansion (°C)$^{-1}$ × 10^{-6}	45–65
	Water absorption after 24 hours at 25°C (%)	0.15–0.17
Chemical properties	Good chemical resistance. Due to their crosslinked or network structures, epoxies can be swollen by polar solvents including acetone, but not dissolved. Tend to yellow and become brittle due to photo-oxidation.	
	Degree of crystallinity (%)	0

Name/Abbreviation	Melamine formaldehyde (MF)	
Repeat unit		
Trade names and synonyms	Formica®	
History	MF was developed in the 1930s and 40s by American Cyanide, Ciba and Henkel and largely replaced urea-formaldehyde.	
Major applications	Radio housings, laminates for kitchen and bathroom surfaces, tableware, cutlery handles.	
Optical properties	Transparent, pale colours available.	
Physical and thermal properties	MF is a thermoset. It is brittle and stiff, so requires fillers or reinforcements. High tensile strength. Low water absorption.	
	Density (g/cm^3)	1.1–1.5
	Maximum use temperature (°C)	80–140
	Ignition time(s)	>300
	Tensile strength at break (MPa)	35–75
	Elongation at break (%)	1.5–2.0
	Coefficient of thermal expansion (°C)$^{-1} \times 10^{-6}$	68
Chemical properties	Highly resistant to dilute acids and alkalis but attacked by concentrated materials.	
	Degree of crystallinity (%)	0

Name/Abbreviation	Phenol-formaldehyde (Bakelite)	
Repeat unit		
Trade names and synonyms	Bakelite, Micarta®, Nomex® (phenolic impregnated paper), Tufnol® (phenolic impregnated textile)	
History	Bakelite was developed in 1907–09. Bakelite Corp. was formed in 1922 and Bakelite Ltd in 1927 in Birmingham, United Kingdom. It was acquired by Union Carbide and Carbon Corp on closing in 1998.	
Major applications	Phenolics can be cast into rod, tubes and sheet and foamed. They can be impregnated into paper and cloth. Electrical insulation, plugs and switches, radio and television housings, telephones, cameras (e.g. Kodak's Brownie 127), fuse box covers, bushings, gears, beads, knife handles, paperweights, billiard balls, saucepan handles and knobs, panelling (as foam).	
	Bakelite is little used today due to cost, complexity of production and its brittleness. It is used for precision-shaped components including saucepan handles and electrical switches.	
Optical properties	Mostly available in dark, opaque colours (novolaks), though transparent, unfilled castings are formed by reacting cresols or resourcinol with formaldehyde (resoles). May be painted or decorated by electroplating.	
	Refractive index (ND 20°C)	1.54–1.58
Physical and thermal properties	Bakelite is a thermoset, hard, brittle material. Its high shrinkage on moulding and brittleness are mediated by adding fillers or reinforcing fibres. It resists burning.	
	Density (g/cm^3)	1.7–2.0
	Ignition time(s)	>300
	Tensile strength at break (MPa)	34–62
	Elongation at break (%)	1.5–2.0
	Coefficient of thermal expansion (°C)$^{-1}$ × 10^{-6}	68
Chemical properties	Resistant to organic liquids but attacked by concentrated nitric and sulphuric acids and chlorine. Photo-oxidation manifests as reduction in gloss of surfaces.	
	Degree of crystallinity (%)	0

Name/Abbreviation	Polyamides (PA), Nylons	
Repeat unit		
Trade names and synonyms	Kevlar®, Perlon®, nylons 46,6,66,610,11 and 12.	
History	Nylon 66 patented in 1931. Large-scale production of Nylon 6 started in 1943.	
Major applications	Woven textiles – both pure and blended, thread, fishing line, rope, brush bristles, hook and loop fasteners (Velcro®), protective clothing, tights and stockings, electrical housings, plumbing connections, bearings, carpeting, reinforcing material, powder coatings.	
Optical properties	Nylons can be prepared either as optically clear or opaque materials. Nylons with higher crystallinities are more opaque than the more amorphous. Some types can be electroplated to allow metallization and most nylons accept printing.	
	Refractive index (ND 20°C)	1.52
Physical and thermal properties for Nylon 6,6 poly (hexamethylene adipamide)	Nylons are tough and strong and can be used over a wide temperature range (-80 to 120°C). Nylons have a low coefficient of friction. This leads to extremely good abrasion resistance, further improved by adding surface lubricants or annealing at 150–200°C. Nylons perform unreliably in wet environments because they absorb water readily.	
	Density (g/cm^3)	1.13–1.15
	Tg (°C)	57
	Tm (°C)	255–265
	Ignition time (s)	10
	Tensile strength at break (MPa)	76–94
	Elongation at break (%)	15–300
	Coefficient of thermal expansion $(°C)^{-1} \times 10^{-6}$	80
	Permeability to oxygen at 30°C $(10^{10} \, cm^3 \, s^{-1} \, mm \, cm^{-2} \, cmHg^{-1})$	0.38
	Permeability to water at 25°C $(10^{10} \, cm^3 \, s^{-1} \, mm \, cm^{-2} \, cmHg^{-1})$	7000
Chemical properties	The numeric suffixes in the names of nylons refer to the number of carbon atoms present in the molecular structures of the start amine and acid respectively (or a single suffix if the amine and acid groups are part of the same molecule).	
	The presence of —CONH— groups causes the difference between the properties of PE and those of nylons.	
	Nylons absorb up to 4% water, so must be conditioned prior to moulding to prevent dimensional changes. Polar nylons including Nylon 46,6,66,610,11 and 12 show high resistance to hydrocarbons, but are swollen or dissolved by alcohols. Nylons are poorly resistant to mineral and organic acids and oxidizing agents but are not affected by alkalis.	
	Degree of crystallinity (%)	30–40

Name/Abbreviation	Polycarbonate (PC)	
Repeat unit		
Trade names and synonyms	Lexan®, Makrolon®, Merlon®, Panlite®, Tuffak®	
History	First synthesized in 1898. Commercially available 1958. Structural foams available in 1973.	
Major applications	Usually shaped by extrusion or thermoforming. Foams used for sports equipment (e.g. water ski shoes) due to high rigidity and fatigue resistance.	
	Compact discs (CDs) and Digital Video Discs (DVD), flat roofs, greenhouse windows, instrument panels, warm and power tool housings, motorcycle helmets, visors for astronauts, bullet-proof and shatter-proof 'glass', riot shields, vehicle bumpers, plates, cups, babies' bottles.	
Optical properties	Highly transparent because PC is amorphous, but thick films cast from solvents are cloudy because slow drying encourages crystallization.	
	Refractive index (ND 20°C)	1.59
	Light transmission (%)	85
Physical and thermal properties	PC has good toughness and rigidity even at elevated temperatures and down to −40°C due to the presence of benzene ring and carbonate (—OCOO—) side groups. It exhibits high impact strength, but tends to craze on loading. PC may be reinforced with glass fibres.	
	Density (g/cm^3)	1.2
	Tg (°C)	150
	Tm (°C)	decomposes above 300
	Ignition time (s)	37
	Tensile strength at break (MPa)	65.5
	Elongation at break (%)	80–150
	Coefficient of thermal expansion (°C)$^{-1} \times 10^{-6}$	68
	Permeability to oxygen at 30°C (10^{10} cm^3 s^{-1} mm cm^{-2} cmHg^{-1})	1
	Permeability to water at 25°C (10^{10} cm^3 s^{-1} mm cm^{-2} cmHg^{-1})	0.04
Chemical properties	Polycarbonate is a generic term for polymers that contain carbon acid esters in the chains of their molecules. Pale yellow colour on manufacture due to impurities in the monomers (bisphenol-A), which is masked with blue dyes. PC is a polar polymer and is dissolved by chlorinated solvents. Almost self-extinguishing on burning. Poor resistance to chemicals and weathering.	
	Degree of crystallinity (%)	0–5%

Name/Abbreviation	Polyethylene (PE)	
Repeat unit		
Trade names and synonyms	Ethylux®, Hitech®, Plastazoate®, polythene	
History	First synthesized in 1933. Process patented 1937.	
Major applications	Produced as film, sheet, rod, fibre and foam.	
	Zip-lock bags, carrier bags, microwave-safe shrink wrap food film, Tupperware® containers, food chopping boards, food packaging films, milk, bread and beer crates, water pipes, electrical cable insulation, geo-textiles, replacement body parts (e.g. heart valves, artificial arteries, hip joints).	
Optical properties	In bulk form, PE is translucent or opaque, but thin films may be transparent. It is difficult to print directly onto PE without pretreatment.	
	Refractive index (ND 20°C)	1.51
	Light transmission (%)	0–75 (LDPE) 0–40 (HDPE)
Physical and thermal properties	PE is thermoplastic. It has a waxy feel and a lower density than water at the same temperature. It is easy to mould, shape and recycle. PE can be readily laminated. It is a more effective thermal insulator than most thermoplastics, therefore requires more energy to raise its temperature. PE is tough with moderate tensile strength. Load-bearing PE deforms continuously with time (creeps). PE has high permeability to gases due to high accessibility through the amorphous areas of the polymer. Permeability to water vapour is low.	
	Density (g/cm^3)	0.92–0.93 (LDPE) 0.95–0.97 (HDPE)
	Tg (°C)	−20 or −110 (LDPE) −90 (HDPE)
	Tm (°C)	110 (LDPE) 137 (HDPE)
	Tensile strength at break (MPa)	8–30 (LDPE) 20–30 (HDPE)
	Elongation at break (%)	100–650 (LDPE) 10–1200 (HDPE)
	Coefficient of thermal expansion (°C)$^{-1}$ × 10^{-6}	100–220 (LDPE) 60–110 (HDPE)
	Ignition time (s)	15 (HDPE)
	Permeability of oxygen at 30°C (10^{10} cm^3 s^{-1} mm cm^{-2} cmHg^{-1})	35–352
	Permeability of water at 25°C (10^{10} cm^3 s^{-1} mm cm^{-2} cmHg^{-1})	130–800

Name/Abbreviation	Polyethylene (PE)	
Chemical properties	Low density (LDPE) and very low density polyethylenes (VLDPE) have branched chains which do not pack together well, imparting low density. High density polyethylene (HDPE) has longer, less branched chains, imparting stiffness and strength. No solvents at ambient temperature due to PE's high crystallinity, but chlorinated liquids can swell the plastic.	
	Stable in contact with alkalis and acids excepting nitric. PE discolours and experiences reduced mechanical properties in the presence of ultraviolet light at ambient temperatures and in the dark above 50°C.	
	Degree of crystallinity (%)	40–50 (LDPE) 70–80 (HDPE)

Name/Abbreviation	Poly (ethylene terephthalate) (PET), polyesters	
Repeat unit		
Trade names and synonyms	Cronar® ,Melinex®, Mylar®, Teijin®, Teonex®, Tetoron®, Dacron®, Terylene®	
History	Polyethylene terephthalate (PET) and the first polyester fibre (Terylene®) were patented in 1941.	
Major applications	Polyesters are available in film, moulded and fibre forms.	
	Fizzy drink bottles, beer bottles, audio and video film, movie film, photographic film, metallized balloons and packaging, credit cards, clothing, carpet, fleece, vehicle body panels, boat hulls, surfboards, roofing, sails, tennis rackets.	
Optical properties	Crystal clear.	
	Refractive index (ND 20°C)	1.52–1.55
	Light transmission (%)	90
Physical and thermal properties	Polyesters are formed via a condensation reaction. PET and poly(butylene terephthalate) (PBT) are not crosslinked so are thermoplastic, but cured polyesters are crosslinked and thermosets. Polyesters may be reinforced with glass or carbon fibres or particles. Addition of 30% or more carbon fibre induces electrical conductivity in polyesters. Despite its high polarity, PET is a good electrical insulator at ambient conditions because its Tg is considerably higher than room temperature. Polyesters have good abrasion resistance and are tough. PET is impervious to water but has low permeability to oxygen.	
	Density (g/cm^3)	1.29–1.40
	Tg (°C)	73–80
	Tm (°C)	245–265
	Ignition time (s)	38
	Tensile strength at break (MPa)	48–72

Name/Abbreviation	Poly (ethylene terephthalate) (PET), polyesters	
	Elongation at break (%)	30–300
	Coefficient of thermal expansion $(°C)^{-1} \times 10^{-6}$	65
	Permeability to oxygen at 30°C $(10^{10} \, cm^3 \, s^{-1} \, mm \, cm^{-2} \, cmHg^{-1})$	0.22
	Permeability to water at 25°C $(10^{10} \, cm^3 \, s^{-1} \, mm \, cm^{-2} \, cmHg^{-1})$	1300
Chemical properties	The ester group (—COO—) imparts susceptibility to hydrolysis and chain scission for polyesters. Effective solvents must interact with the ester groups. Chlorinated solvents are most frequently used.	
	Degree of crystallinity (%) (up to 42% when film is annealed under heating and restraint)	0–30

Name/Abbreviation	Polymethylmethacrylate (PMMA), acrylic	
Repeat unit		
Trade names and synonyms	Acrylite®, Acrylplast®, Altuglas®, Lucite®, Perspex®, Plexiglas®, Plazcryl®, poly (methyl 2-methylpropenoate) (IUPAC name)	
History	Developed in 1872 and commercialized in 1933 as cockpit canopies for fighter planes in World War II.	
Major applications	Available as sheet, rod or tube and fibres. Shaped by thermoforming, casting or extrusion.	
	Lenses for spectacles and optical equipment, contact lenses, aircraft windows, greenhouse windows, cockpit canopies, safety spectacles, information and advertising signs, food containers, aquaria, domestic and vehicle lighting, furniture, electrical components, artificial fingernails.	
Optical properties	Acrylics are highly transparent and replace glass in many applications because the polymer is amorphous. It scratches readily but may be coated to increase its resistance to damage. If scratched, polishing can restore the original appearance but with risk of stress cracking. Because the critical angle for PMMA/air boundary is 42°, light may be totally internally reflected through long lengths of solid plastic, also through curved areas. This principle allows the application of PMMA in optical fibres.	
	Refractive index (ND 20°C)	1.49
	Light transmission (%)	92

Name/Abbreviation	Poly (methyl methacrylate) (PMMA), acrylic	
Physical and thermal properties	PMMA is thermoplastic, hard and rigid. It is tougher than PS but less so than cellulose acetate or ABS. It is impermeable to oxygen, so contact lenses should only be worn for short periods to allow the eye access to oxygen. PMMA's wetting angle is around 18°, so tears are able to wet surfaces of PMMA lenses allowing lenses to move freely over the eye. PMMA may be blended with PVC to increase durability and toughness.	
	Density (g/cm^3)	1.17–1.20
	Tg (°C)	85–105
	Tm (°C)	decomposes above 160
	Tensile strength at break (MPa)	48–76
	Elongation at break (%)	2–10
	Coefficient of thermal expansion (°C)$^{-1}$ × 10^{-6}	50–90
	Permeability to oxygen at 30°C (10^{10} cm^3 s^{-1} mm cm^{-2} cmHg^{-1})	0.5
	Water absorption (% in 24 h at 20°C)	0.2–0.3
Chemical properties	The bulky shape of the MMA group ensures an amorphous structure because it prevents close packing of polymer chains. PMMA has extremely good weathering resistance. Acrylics are dissolved or stress crazed on contact with ketones, esters, hydrocarbons, acids and alkalis.	
	Degree of crystallinity (%)	0

Name/Abbreviation	Polypropylene (PP)		
Repeat unit	$-\overset{\displaystyle H}{\underset{\displaystyle H}{C}}-\overset{\displaystyle H}{\underset{\displaystyle CH_3}{C}}-$		
Trade names and synonyms	Rubbermaid®, Sterilite®, polypropene, propene polymers		
History	First produced commercially in 1958.		
Major applications	Produced as film, rod, foam and fibre.		
	Food containers, domestic kettles, pipes, ropes, cable insulation, car bumpers and dashboards, artificial grass, furniture shells, crates, thermal clothing, medical instruments, plastic banknotes.		
Optical properties	Unpigmented PP is more transparent than PE, despite their similar crystallinities, due to lower crystal density of the former.		
	Refractive index (ND 20°C)	1.49	
	Light transmission (%)	55–90	

Name/Abbreviation	Polypropylene (PP)	
Physical and thermal properties	PP is thermoplastic. It has low density and low strength but has poor impact resistance. Reinforcing with glass, chalk or talc increases flexibility, impact resistance and strength. PP has a higher melting temperature than PE so can be used to contain hot materials and to withstand steam sterilizing treatments.	
	Density (g/cm^3)	0.90–0.91
	Tg (°C)	−20
	Tm (°C)	168–175
	Degradation temperature (°C)	286
	Ignition time (s)	17
	Tensile strength at break (MPa)	31–41
	Elongation at break (%)	100–600
	Coefficient of thermal expansion (°C)$^{-1}$ × 10^{-6}	80–100
	Permeability to oxygen at 30°C (10^{10} cm^3 s^{-1} mm cm^{-2} cmHg^{-1})	23
	Permeability to water at 25°C (10^{10} cm^3 s^{-1} mm cm^{-2} cmHg^{-1})	680
Chemical properties	PP occurs in isotactic (all methyl groups on one side of the molecule, syndiotactic (methyl groups on alternating sides) and atactic (random placing of methyl groups) stereo-regular forms. Atactic PP is amorphous and rubbery while isotactic material is stiff and highly crystalline. Commercial PP is 90–95% isotactic. Increased strength and flexibility are obtained by copolymerizing propylene with ethylene or by incorporating reinforcing additives such as rubber and glass fibres.	
	PP is more susceptible than PE to oxidation due to the presence of a reactive tertiary carbon atom. It is not vulnerable to environmental stress cracking. PP has no solvents at ambient temperature due to its high crystallinity. PP is highly resistant to attack by alkali and acids excepting concentrated sulphuric and chromic acids.	
	Degree of crystallinity (%)	50–60

Name/Abbreviation	Polystyrene (PS)
Repeat unit	
Trade names and synonyms	Solid: Stryroloxyld®, Nunc®, Costar®
	Foam: Styrafoam®, expanded polystyrene (EPS), flamingo, Fome-Cor®

Name/Abbreviation	Polystyrene (PS)	
History	First synthesized in 1839. Mass production developed ca.1931.	
Major applications	Produced as film, moulding and foam (approximately one third of commercial PS is foamed).	
	Disposable cups and glasses for hot (only PS foam) and cold drinks, electrical fittings, CD case (jewel box), Airfix® model kits, food containers, coat hangers, hair combs, TV cabinets, expanded foam packaging, foam thermal insulation, foam ceiling tiles, insulated food containers.	
Optical properties	Polystyrene's high clarity is due to its amorphous structure. Its high transmission of all wavelengths in visible light and high refractive index impart a glass-like sparkle. It is used when appearance and low cost are high priority factors and mechanical loading requirements are low. It can haze and yellow with time due to impurities in the styrene monomer.	
	Refractive index (ND 20°C)	1.59
	Light transmission (%)	87–92
Physical and thermal properties	PS is thermoplastic and burns with a sooty flame. PS is brittle and cannot withstand boiling water. It emits a characteristic metallic ring when dropped. Its mechanical properties can be improved dramatically by blending it with polybutadiene. Addition of even small amounts of plasticizer results in unacceptable reduction in softening temperature, which limits the usefulness of the plastic. PS's pure hydrocarbon structure and low water absorption result in excellent thermal and electrical insulation properties which are retained in wet conditions.	
	Density (g/cm^3)	1.04–1.05
	Tg (°C)	74–105
	Tm (°C)	100–108
	Ignition time (s)	15
	Tensile strength at break (MPa)	36–52
	Elongation at break (%)	1.2–2.5
	Coefficient of thermal expansion (°C)$^{-1} \times 10^{-6}$	50–83
	Permeability to oxygen at 30°C (10^{10} cm^3 s^{-1} mm cm^{-2} cmHg^{-1})	11
	Permeability to water at 25°C (10^{10} cm^3 s^{-1} mm cm^{-2} cmHg^{-1})	1200
Chemical properties	PS is amorphous because its benzene ring, when present on alternate sides of the polymer chain (atactic form), prevents close packing of the polymer chains, which would allow crystallinity. The benzene ring imparts greater reactivity to PS than PE. PS is subject to oxidation resulting in yellowing, particularly in the presence of sulphur. PS is dissolved by hydrocarbons, chlorinated hydrocarbons, ketones, esters and some oils. Acids, alcohols, cosmetic creams and foodstuffs cause crazing and cracking.	
	Degree of crystallinity (%)	0

Name/Abbreviation	Acrylonitrile-butadiene-styrene (ABS)	
Repeat unit		
Trade names and synonyms	Absylux®, Cycolac®, Terulan®	
History	First became available in the 1950s to replace acrylonitrile-styrene copolymers.	
Major applications	Computer cases, telephones, casings for domestic electrical equipment, safety helmets, plastic baths and shower trays, Lego® bricks.	
Optical properties	Usually opaque but may be transparent. ABS has a good surface finish with high gloss.	
	Refractive index (ND 20°C)	1.56–1.57
	Light transmission (%)	87–92
Physical and thermal properties	The acrylonitrile component imparts thermal and chemical resistance, rubbery butadiene imparts ductility and strength while styrene gives a glossy surface and ease of shaping. ABS has the highest impact resistance of all polymers, particularly at low temperatures. ABS has high dimensional stability at elevated temperatures. It has poor flame resistance and can be alloyed with PVC to improve toughness and fire retarding properties.	
	Density (g/cm^3)	1.04–1.07
	Tg (°C)	90–120
	Tm (°C)	106–130
	Ignition time (s)	10
	Tensile strength at break (MPa)	38–48
	Coefficient of thermal expansion (°C)$^{-1}$ × 10^{-6}	85–234
	Permeability to water at 25°C	absorbs up to 0.3% by weight in 24 hours at 25°C
Chemical properties	Produced by adding styrene and acrylonitrile to polybutadiene latex and adding an initiator to polymerize the styrene and acrylonitrile. Ratios of the three main ingredients affect the properties of the copolymer. ABS is highly hygroscopic. It has good chemical resistance and high resistance to stress cracking. ABS weathers poorly so is not suitable for long-term outdoor use.	
	Degree of crystallinity (%)	0

Name/Abbreviation	Polytetrafluoroethylene (PTFE)	
Repeat unit	$$-\overset{\displaystyle F}{\underset{\displaystyle F}{C}}-\overset{\displaystyle F}{\underset{\displaystyle F}{C}}-$$	
Trade names and synonyms	Teflon®, GoreTex®	
History	First synthesized in 1938. Commercialized in late 1940s.	
Major applications	Fibres, textiles (e.g. GoreTex), coatings.	
	Non-stick cookware, water- and bacteria- repellant textiles, wire and cable covers, corrosion-resistant linings for pipes and valves, seals and gaskets, skis.	
Optical properties	White and translucent due to high crystallinity. Limited range of colours available.	
	Refractive index (ND 20°C)	1.35
Physical and thermal properties	PTFE products are thermoplastic and can be used between −270°C and 250°C.	
	PTFE has extremely low friction and very high resistance to wetting. Soft, waxy feel partly due to low coefficient of friction. Can be reinforced with glass, nylon or Kevlar® polyamide fibres to produce a skin of high toughness, strength and water-resistance. PTFE has very low permeability to gases and water vapour.	
	Density (g/cm^3)	2.14–2.20
	Tg (°C)	−90
	Tm (°C)	327
	Tensile strength at break (MPa)	14–34
	Elongation at break (%)	200–300
	Coefficient of thermal expansion (°C)$^{-1} \times 10^{-6}$	70–120
	Permeability to oxygen at 30°C (10^{10} cm^3 s^{-1} mm cm^{-2} cmHg^{-1})	0.1
	Permeability to water at 25°C (10^{10} cm^3 s^{-1} mm cm^{-2} cmHg^{-1})	2.9
Chemical properties	PTFE is a linear polymer, free from branching. PTFE requires very high processing temperatures, around 380°C, which excludes the use of many pigments and additives. Inorganic cadmium compounds, iron oxides and ultramarines may be used. High chemical and solvent resistance. Epoxy, nitrile-phenolic and silicone adhesives can be used.	
	Degree of crystallinity (%)	50–70

Name/Abbreviation	Polyurethane (PUR)	
Repeat unit	$$\begin{array}{ccccccccc} & O & H & & H & O & & O & H & & H & O \\ & \| & \| & & \| & \| & & \| & \| & & \| & \| \\ -O-C-N-R-N-C-O-R'-O-C-N-R-N-C-O- \end{array}$$	
Trade names and synonyms	Bayflex®, Lycra®, Spandex®	
History	Otto Bayer patented PUR adhesives and elastomers in 1937. Mass production of flexible foams for furniture and rigid foams in insulation took place in the 1960s.	
Major applications	Available as fibres, textiles, rubbers, rigid and flexible foams and surface coatings.	
Optical properties	PURs are usually shaped by casting, producing a high quality surface finish and the opportunity for intricate shapes.	
	Refractive index (ND 20°C)	1.6
Physical and thermal properties	Polyurethane polymers, the products of a reaction between polyisocyanates and either polyether or polyester, may be either thermoplastic or thermosetting. Most PUR foams are thermosets. Polyester-based foams have higher mechanical properties than polyether types. Excellent abrasion resistance, flexibility and resistance to cutting. For load-bearing applications, including conveyor belts, PURs are reinforced with nylon fibres to give flexibility and high strength.	
	Density (g/cm^3)	1.02–1.25
	Service temperature range (°C)	−55–120
	Tensile strength at break (MPa)	5.5–55
	Maximum elongation (%)	250–800
Chemical properties	Polyether-based materials have higher resistance to hydrolysis but are more vulnerable to photo-oxidation than polyester types. Good chemical resistance. Solvents cause reversible swelling but not solution. PUs can be adhered with polyurethanes, nitriles, epoxies and cyanoacrylates.	
	Degree of crystallinity (%)	5–10

Name/Abbreviation	Poly (vinyl chloride) (PVC)
Repeat unit	$$\begin{array}{cc} H & H \\ \| & \| \\ -C - C - \\ \| & \| \\ H & Cl \end{array}$$
Trade names and synonyms	Excelon®, Tedlar®, Trovicel®, Tygon®, Verilon®, polychloroethene (IUPAC name), uPVC, PPVC, rigid PVC, vinyl
History	PVC accidentally discovered both in 1835 and 1872. A technique to plasticize PVC was developed in 1926.

Name/Abbreviation	Polyvinyl chloride (PVC)	
Major applications	Available as film, foam, tubes, pipes, thermoform, blow moulding and extruded mouldings.	
	Rigid PVC: window frames and sills, water pipe, gutter, replacement for wood.	
	Plasticized PVC: flexible tubing, garden hoses, vinyl flooring, shower curtains, roofing membranes, electrical cable insulation, waterproof and protective clothing, vehicle upholstery, medical devices, blood bags, intravenous bags and tubing.	
Optical properties	Transparent but not as glass-clear as acrylics or polyesters.	
	Refractive index (ND 20°C)	1.52–1.55
	Light transmission (%)	76–82
Physical and thermal properties	PVC is thermoplastic. In rigid, unplasticized form, PVC is stiff and brittle. Addition of external plasticizer imparts various levels of softness and flexibility. Copolymerization with flexible monomers also results in a softer product. Vinyl records or LPs comprise vinyl chloride/vinyl acetate copolymer.	
	PVC does not burn. In a fire, PVC cables form hydrogen chloride fumes as the major degradation product. The fumes scavenge free radicals thereby preventing the plastic burning.	
	Density (g/cm^3)	1.30–1.58
	Tg (°C)	75–105
	Tm (°C)	212
	Ignition time (s)	300 (unplasticized) 22 (plasticized)
	Tensile strength at break (MPa)	41–52(unplasticized) 19 (35% plasticized)
	Elongation at break (%)	40–80 (unplasticized) 270 (35% plasticized)
	Coefficient of thermal expansion (°C)$^{-1}$ × 10^{-6}	50–100
	Permeability of oxygen at 30°C (10^{10} cm^3 s^{-1} mm cm^{-2} cmHg^{-1})	1.2
	Permeability to water at 25°C (10^{10} cm^3 s^{-1} mm cm^{-2} cmHg^{-1})	1560
Chemical properties	PVC undergoes dehydrochlorination (production of hydrogen chloride gas) under exposure to heat and light. The acidic gas corrodes metals in the vicinity. The reaction is autocatalytic. PVC becomes darker in colour as degradation progresses. Plasticizers are usually esters and undergo hydrolysis in acidic and basic environments to form white crystals of acid or anhydride. The PVC polymer is resistant to acids and bases. It is resistant to most hydrocarbons, although polar solvents may extract plasticizer if present. Water swells the PVC polymer imparting an opacity which reverses on drying. PVC can be adhered using polyester, epoxy or polyurethane adhesives.	
	Degree of crystallinity (%)	5–15%

Name/Abbreviation	Urea-formaldehyde (UF)	
Repeat unit		
Trade names and synonyms	Beetle	
History	Hans John made UF without a catalyst and patented the process in 1920. The first commercial successful UF occurred in 1928 under British Cyanide Co. I.G. Farben took out more patents 1925–28.	
Major applications	Electrical fittings, telephone handsets, radio housings, cigarette boxes, lampshades, tableware, cavity wall insulation (UFFI) from 1950s until 1980 when it was discontinued due to concern over formaldehyde toxicity.	
Optical properties	Transparent, pale colours available.	
Physical and thermal properties	UF is a thermoset. It is brittle and stiff, so requires filler or reinforcement. Organic fillers are required for UF because it is more sensitive to the high temperatures required for inorganic fillers than MF. High tensile strength. Low water absorption.	
	Density (g/cm^3)	1.1–1.5
	Maximum use temperature (°C)	80–140
	Ignition time (s)	>300
	Tensile strength at break (MPa)	35–75
	Elongation at break (%)	1.5–2.0
	Coefficient of thermal expansion (°C)$^{-1}$ × 10^{-6}	68
Chemical properties	Highly resistant to dilute acids and alkalis but attacked by concentrated materials.	
	Degree of crystallinity (%)	0

References

Brydson, J. A. (1999). *Plastics Materials*, 6th edition. Butterworth-Heinemann.

Pedersen, L. B. (1999). *Plast org Miljø (Plastics and Environment, in Danish)*. Teknisk ForlagA/S, Copenhagen.

Appendix 2
Fourier Transform Infrared Spectra of polymers frequently found in collections (arranged in alphabetical order)

Cellulose acetate

Source of sample: Photographic film without emulsion from the 1950s
Sample preparation: None. ATR accessory used to run FTIR spectrum

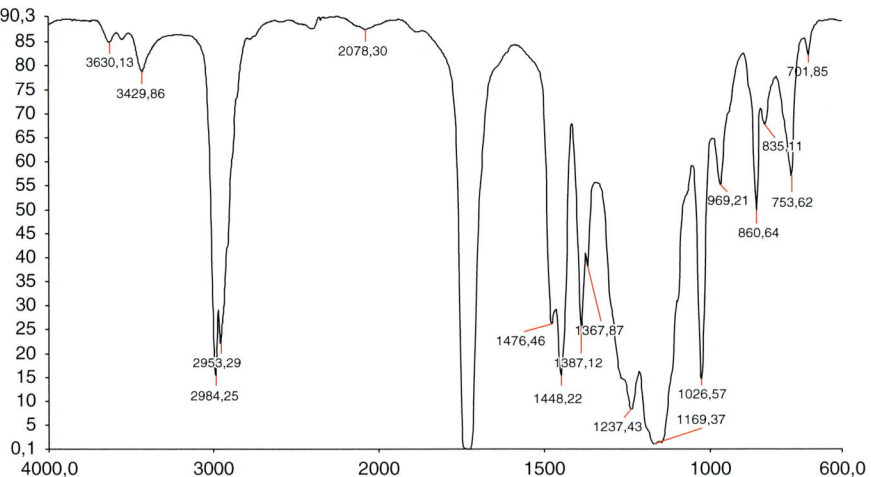

Table A2.1 Characteristic peaks for cellulose acetate		
Wavenumber (cm^{-1})	**Type of vibration**	**Assignment**
3430	stretching	OH..O
3000–2840	asymmetric and symmetric stretching	CH_2 and CH_3
1743	stretching	C=O
1368	in-plane deformation	CH_3
1237	stretching	C=O and C—O
1027	stretching	C—O

Cellulose nitrate

Source of sample: Degraded spectacle frames from the 1920s
Sample preparation: None. ATR accessory used to run FTIR spectrum

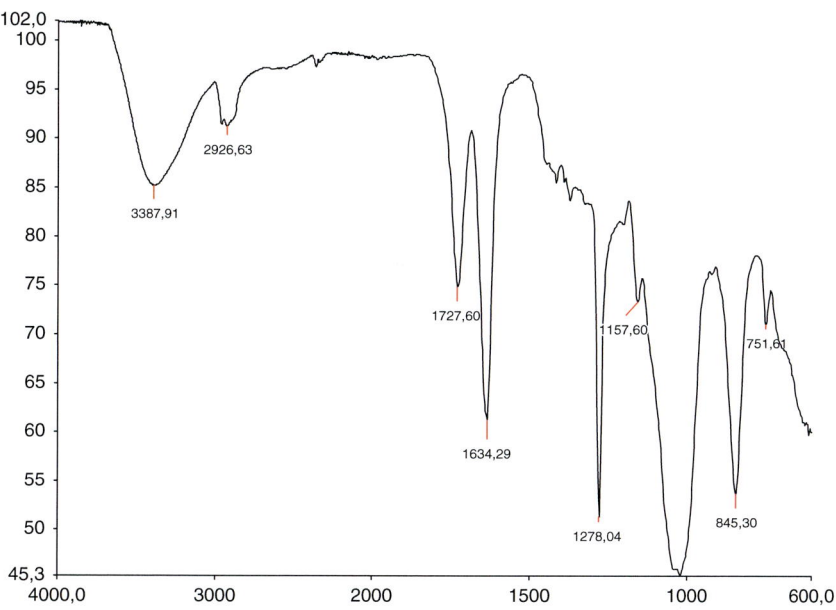

Table A2.2 Characteristic peaks for cellulose nitrate		
Wavenumber (cm^{-1})	**Type of vibration**	**Assignment**
3389	stretching	OH..O
3000–2840	asymmetric and symmetric stretching	CH_2
1634	asymmetric stretching	NO_2
1278	symmetric stretching	NO_2
1063	stretching	C—O
845	rocking vibration	CH_2

Epoxy

Source of sample: Bisphenol A-based casting epoxy
Sample preparation: None. ATR accessory used to run FTIR spectrum

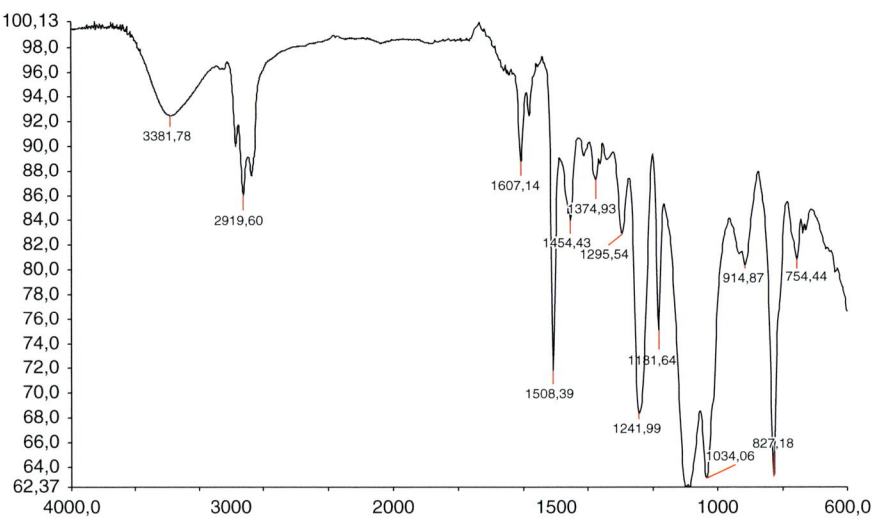

Table A2.3 Characteristic peaks for epoxy		
Wavenumber (cm⁻¹)	**Type of vibration**	**Assignment**
3382	stretching	OH..O
3000–2840	asymmetric and symmetric stretching	CH_2 and CH_3
1610–1325	stretching	benzene ring
1454	in-plane deformation	CH_2 and CH_3
1374	in-plane deformation	OH and CH_3
1296	in-plane deformation	CH
1242	asymmetric stretching	C—O—C
1182 and 1106	in-plane deformation	CH
1034	stretching and rocking	C—O and CH_3
827	rocking	CH_2

Phenol-formaldehyde (phenolic)

Source of sample: Unfilled cast made of a novolak with colour added

Sample preparation: None. ATR accessory used to run FTIR spectrum

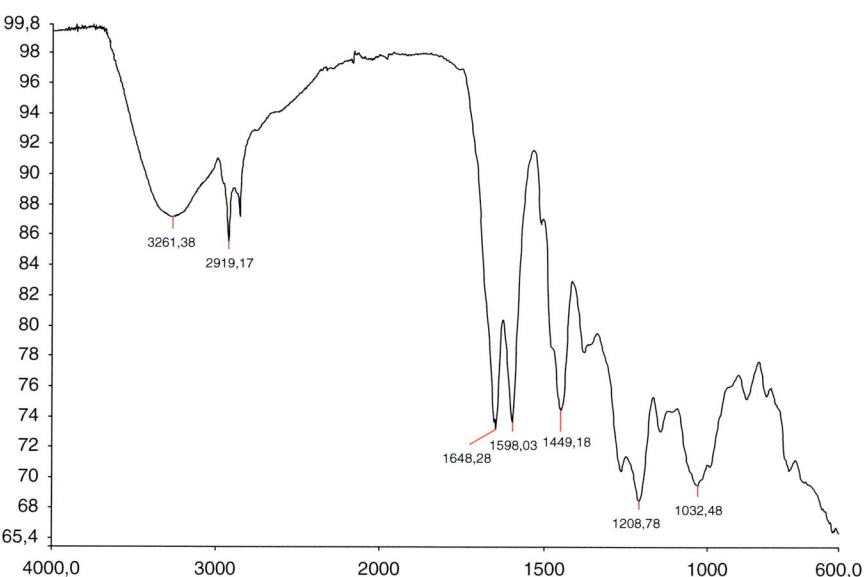

Wavenumber (cm^{-1})	Type of vibration	Assignment
	Table A2.4 Characteristic peaks for phenol-formaldehyde	
3261	stretching	OH..O
3150–3000	stretching	=CH
3000–2840	asymmetric and symmetric stretching	CH$_2$
1600–1320	stretching	benzene ring
1450	in-plane deformation	CH$_2$
1209	in-plane deformation	benzene ring-O
1095	stretching	C—O

Polyamide (nylon)

Source of sample: Nylon curtain ring

Sample preparation: None. ATR accessory used to run FTIR spectrum

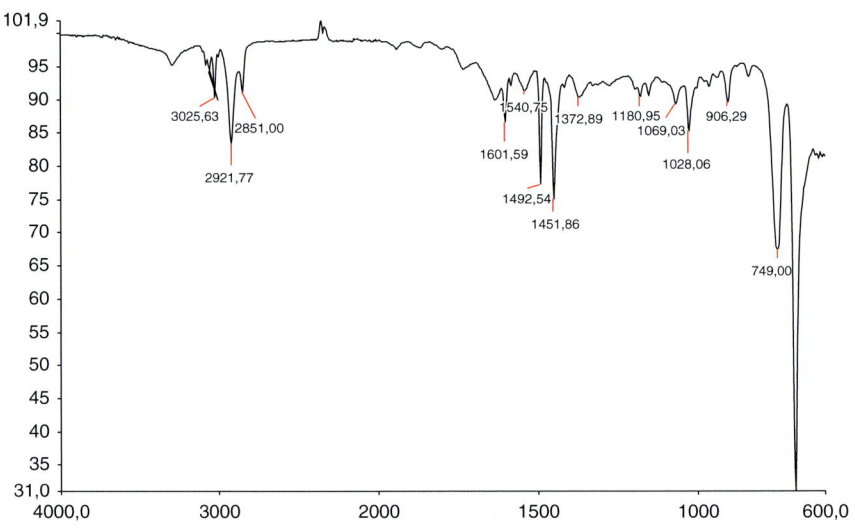

Table A2.5 Characteristic peaks for nylon		
Wavenumber (cm^{-1})	**Type of vibration**	**Assignment**
3300	stretching	N—H
2922	asymmetric stretching	CH$_2$
2851	symmetric stretching	CH$_2$
1602	stretching	C=O
1541	in-plane deformation	NH
	stretching	CN
1275	in-plane deformation	NH
	stretching	CN
1200	wagging and twisting	CH$_2$
700	out-of-plane deformation	NH and C=O

Polycarbonate

Source of sample: CD without metallized layer or coating
Sample preparation: None. ATR accessory used to run FTIR spectrum

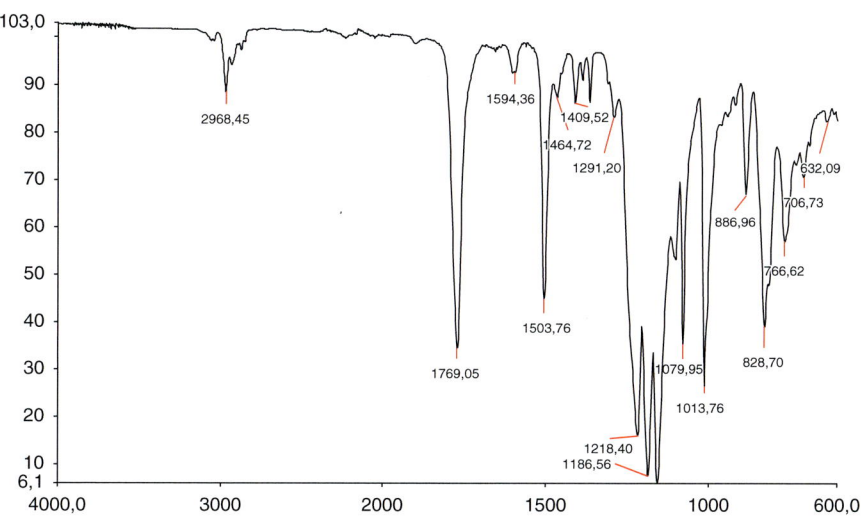

Table A2.6 Characteristic peaks for polycarbonate		
Wavenumber (cm^{-1})	**Type of vibration**	**Assignment**
3150–3000	stretching	=CH
3000–2840	asymmetric and symmetric stretching	CH$_3$
1769	stretching	C=O
1504	stretching	benzene ring
1218	asymmetric stretching	C—O—C
1186	in-plane deformation	=CH
1162	in-plane deformation	=CH
1015	in-plane deformation	=CH
829	out-of-plane deformation	=CH

Polyethylene

Source of sample: Low density polyethylene zip-lock bag
Sample preparation: None. ATR accessory used to run FTIR spectrum

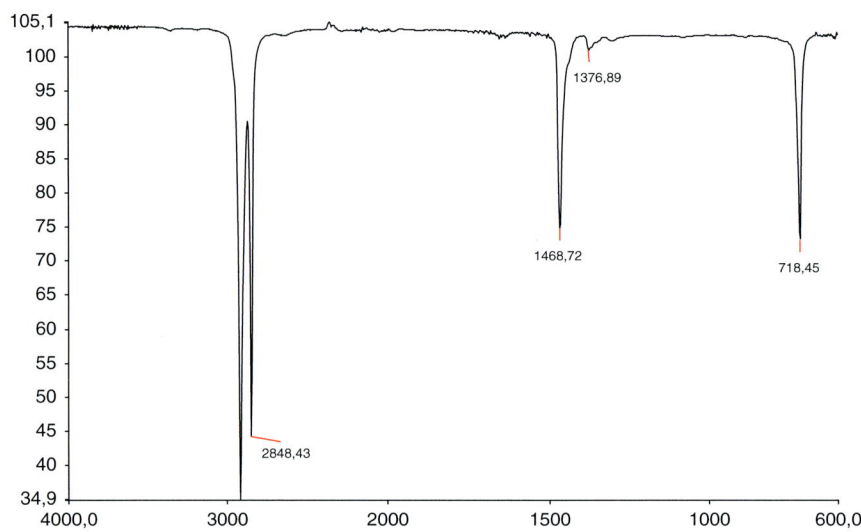

Table A2.7 Characteristic peaks for polyethylene		
Wavenumber (cm^{-1})	**Type of vibration**	**Assignment**
3000–2840	asymmetric and symmetric stretching	CH_2
1469	in-plane deformation	CH_2
718	rocking vibration	CH_2

Polyethylene terephthalate (saturated polyester)

Source of sample: Fizzy drink bottle
Sample preparation: None. ATR accessory used to run FTIR spectrum

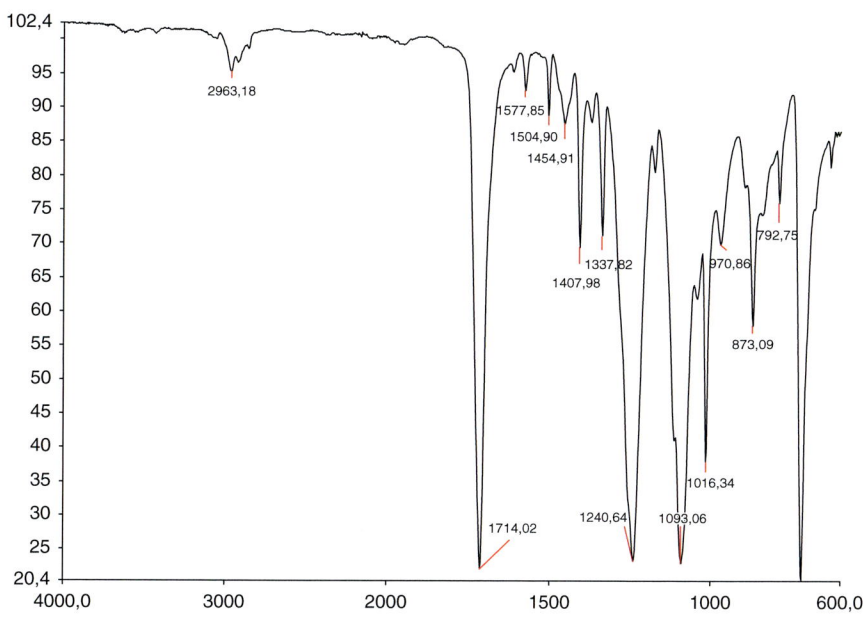

Table A2.8 Characteristic peaks for polyesters		
Wavenumber (cm^{-1})	**Type of vibration**	**Assignment**
3150–3000	stretching	CH
3000–2840	asymmetric and symmetric stretching	CH$_2$
1714	stretching	C=O
1600–1325	stretching	benzene ring
1240	stretching	C=O and C—O
	in-plane deformation	=CH
1093	stretching	O—C
	in-plane deformation	=CH
1016	in-plane deformation	=CH
971	out-of-plane deformation	=CH
873	out-of-plane deformation	=CH
726	out-of-plane deformation	benzene ring

Poly (methyl methacrylate) (acrylic)

Source of sample: Transparent Plexiglas sheet
Sample preparation: None. ATR accessory used to run FTIR spectrum

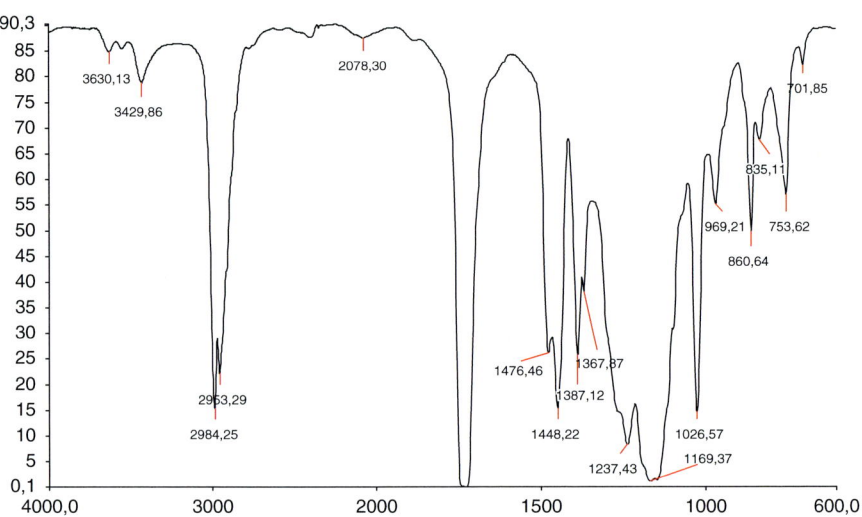

Table A2.9 Characteristic peaks for poly (methyl methacrylate)		
Wavenumber (cm^{-1})	**Type of vibration**	**Assignment**
3000–2840	asymmetric and symmetric stretching	CH_2 and CH_3
1734	stretching	$C{=}O$
1476	in-plane deformation	CH_2 and CH_3
1387	in-plane deformation	CH_3
1238	stretching	$C{=}O$ and $C{-}O$
1169	rocking vibration	CH_3
970	stretching	$C{-}O$
754	in-plane deformation	$C{=}O$
	rocking vibration	CH_2

Polypropylene

Source of sample: Disposable fo od packaging
Sample preparation: None. ATR accessory used to run FTIR spectrum

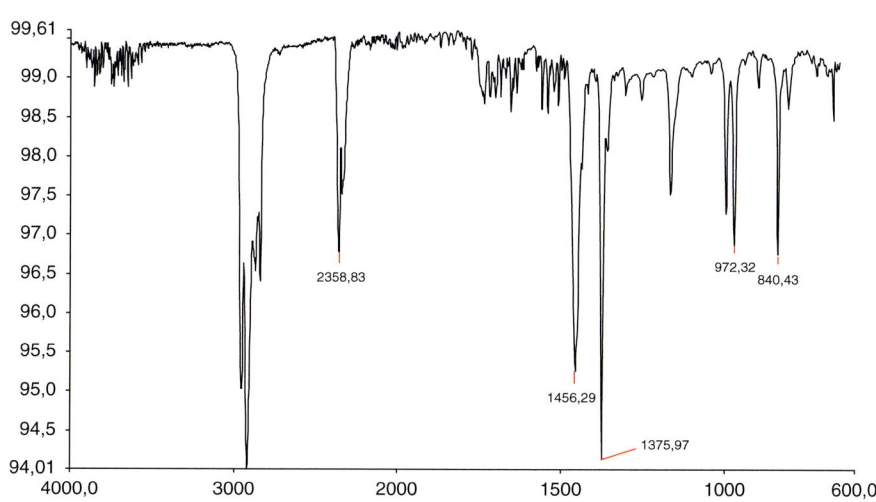

Table A2.10 Characteristic peaks for polypropylene		
Wavenumber (cm^{-1})	**Type of vibration**	**Assignment**
3000–2840	asymmetric and symmetric stretching	CH_2 and CH_3
1456	in-plane deformation	CH_2 and CH_3
1376	in-plane deformation	CH_3
1167	rocking vibration	CH_3
1000	CH_3	CH_3
972	CH_3	
840	rocking vibration	CH_2

Polystyrene

Source of sample: Polystyrene standard for FTIR spectroscopy
Sample preparation: None. ATR accessory used to run FTIR spectrum

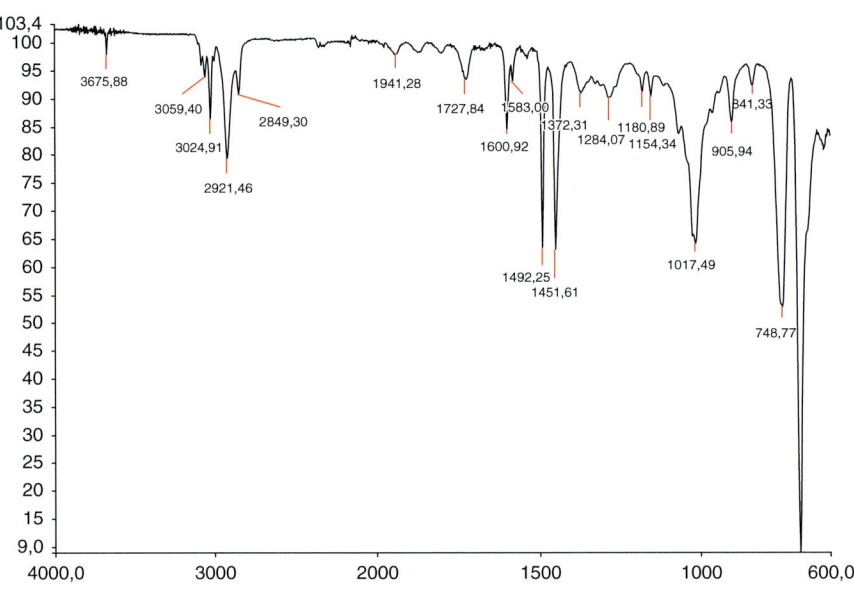

Table A2.11 Characteristic peaks for polystyrene		
Wavenumber (cm⁻¹)	**Type of vibration**	**Assignment**
3150–3000	stretching	$=CH$
3000–2850	asymmetric and symmetric stretching	CH_2
1600–1372	stretching	benzene ring
1181	in-plane deformation	$=CH$
1154	in-plane deformation	$=CH$
1069	in-plane deformation	$=CH$
1028	in-plane deformation	$=CH$
906	out-of-plane deformation	$=CH$
841	out-of-plane deformation	$=CH$
749	out-of-plane deformation	$=CH$

Polytetrafuoroethylene (Teflon)

Source of sample: New septum from gas chromatograph
Sample preparation: None. ATR accessory used to run FTIR spectrum

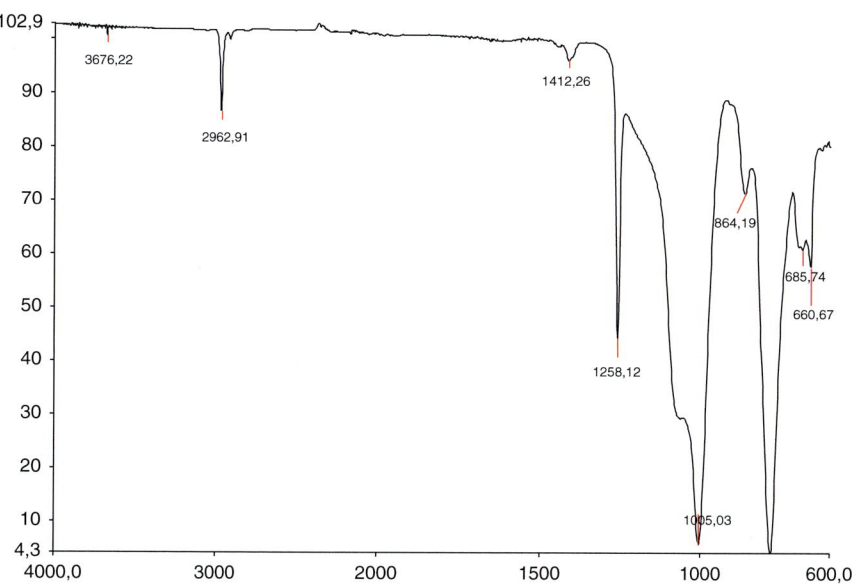

Table A2.12 Characteristic peaks for polytetrafluoroethylene		
Wavenumber (cm^{-1})	**Type of vibration**	**Assignment**
1220	stretching	CF_2
1197	stretching	CF_2
661	in-plane deformation	CF_2

Polyurethane ester

Source of sample: 'LIVESTRONG' arm band
Sample preparation: None. ATR accessory used to run FTIR spectrum

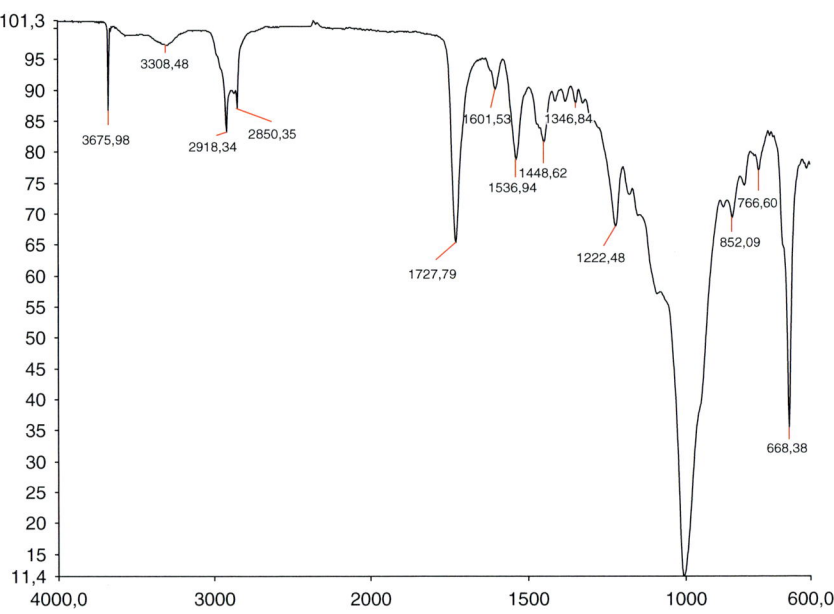

Table A2.13	Characteristic peaks for polyurethane ester	
Wavenumber (cm^{-1})	**Type of vibration**	**Assignment**
3309	stretching	NH
3000–2850	asymmetric and symmetric stretching	CH$_2$
1728	stretching	C=O
1600–1450	stretching	benzene ring
1537	in-plane deformation	NH
1449	in-plane deformation	CH$_2$
1050	stretching	C—O—C

Poly (vinyl chloride) – unplasticized

Source of sample: PVC powder

Sample preparation: None. ATR accessory used to run FTIR spectrum

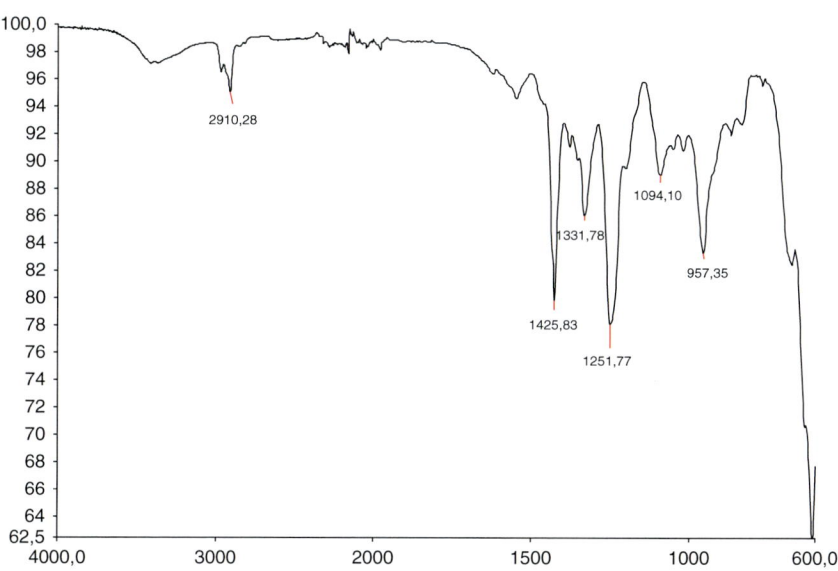

Table A2.14   Characteristic peaks for unplasticized poly (vinyl chloride)

Wavenumber (cm^{-1})	Type of vibration	Assignment
3000–2840	stretching	CH_2 and CH
1426	in-plane deformation	CH_2
1332	wagging vibration	CH
1252	in-plane deformation	CH
958	stretching	C—C
876	rocking vibration	CH_2
615	stretching	C—Cl

Poly (vinyl chloride) – plasticized

Source of sample: PVC photograph pocket with phthalate plasticizer
Sample preparation: None. ATR accessory used to run FTIR spectrum

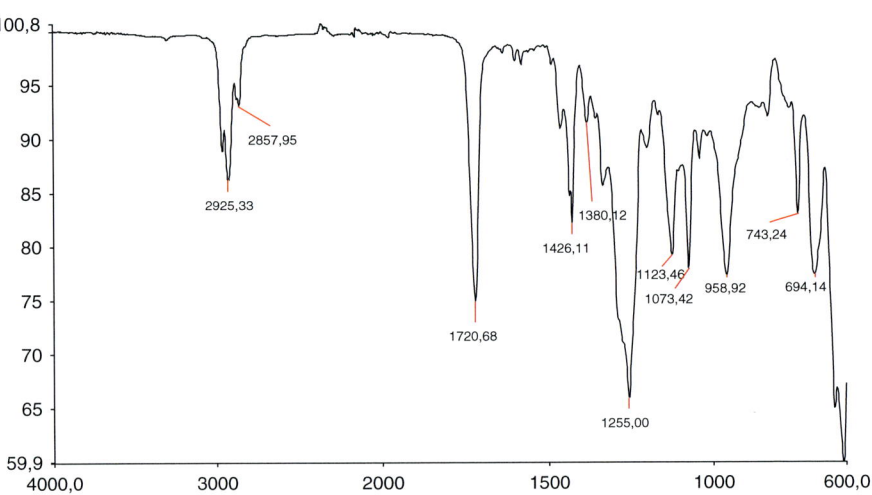

Table A2.15 Characteristic peaks for plasticized poly (vinyl chloride) (peaks in italics are assigned to phthalate plasticiser)

Wavenumber (cm⁻¹)	Type of vibration	Assignment
3000–2840	stretching	CH_2 and CH
1720	*stretching*	*C=O*
1464	*stretching*	*CH*
1426	in-plane deformation	CH_2
1380	*stretching*	*CH*
1332	wagging vibration	CH
1258	*stretching*	*C—O*
1252	in-plane deformation	CH
1123	*stretching*	*C—O*
1073	*stretching*	*C—O*
958	stretching	C—C
876	rocking vibration	CH_2
743	*stretching*	*CH*
615	stretching	C—Cl

Appendix 3 Terms used to describe the degradation of plastics

Degradation term	Description	Appearance	Cause	Plastics affected
abrasion	group of scratches or roughening		physical damage – possibly due to use	all
bloom	white or grey matte film covering surfaces		migration of antioxidants to surfaces	all but mainly synthetic rubbers

Degradation term	Description	Appearance	Cause	Plastics affected
break	separation of object into two or more pieces		physical damage perhaps due to use	all plastics in solid, foam or film form
brittleness	friable or weak surface, easily broken if touched		reduction in molecular weight due to oxidation or hydrolysis	all
chalking	white powder on surfaces		separation of pigment or filler from polymer	all
chip	small missing piece from surface; dip in surface		physical damage – perhaps due to use	hard plastics, e.g. polystyrene, acrylics, polycarbonate
crack	line of failure which does not penetrate entire thickness		physical damage – perhaps due to use	hard plastics, e.g. polystyrene, acrylics, polycarbonate

Degradation term	Description	Appearance	Cause	Plastics affected
craze	interconnecting cracks at surfaces or in body of object		hydrolysis of semi-synthetic plastics or stress crazing by polar solvents	cellulose nitrate and acetate, polystyrene
crumbling	loss of coherence in material resulting in powdering		large reduction in molecular weight due to polymer chains breaking	polyurethane and polyethylene foams
dent	deformation, either permanent or temporary		physical damage – perhaps due to use; thermal damage from overheating	thermoplastics, e.g. polyethylene, PVC
discolouration	change in colour of opaque plastics		formation of chromophores due to light or change in appearance of colouring agents, e.g. dyes	all filled and coloured plastics
pitting	craters or pits on surfaces *not* caused by mechanical damage		evaporation of plasticizers and solvents during curing or shaping	plasticized PVC, epoxy, acrylics
scratch	indentation made mechanically		physical damage – perhaps due to use or poor handling	all plastics

Degradation term	Description	Appearance	Cause	Plastics affected
stain	foreign marks which have been transferred to plastics		labelling materials or close contact with printed/written paper	plasticized PVC, e.g. document wallets
shrinkage	material occupies less space than originally, deforms		loss of plasticizer, other additives or crosslinking	all plastics
sweating	beads of liquid on surfaces		migration of plasticizer or other liquid additives from polymer chains	plasticized PVC, polyethylene
yellowing	darkening or formation of yellow tones in white or colourless plastics only		formation of chromophores	all

INDEX